A Companion to Interdisciplinary STEM Project-Based Learning

A Companion to Interdisciplinary STEM Project-Based Learning

For Teachers by Teachers

Edited by

Robert M. Capraro
Mary Margaret Capraro
Jim Morgan
Jim Scheurich

Texas A&M University, USA

SENSE PUBLISHERS
ROTTERDAM/BOSTON/TAIPEI

A C.I.P. record for this book is available from the Library of Congress.

ISBN:78-94-6091-117-0 (paperback)
ISBN:978-94-6091-119-4 (e-book)

Published by: Sense Publishers,
P.O. Box 21858,
3001 AW Rotterdam,
The Netherlands
http://www.sensepublishers.com

Printed on acid-free paper

DEDICATION

We at the Aggie STEM Center are glad you have an interest in this book and the work of the enthusiastic and very dedicated STEM teachers who contributed while tirelessly striving to improve STEM education in some of the most challenging urban high schools.

I would like you to consider three things as you read this book. First, this book was assembled by teachers, for teachers. It represents lessons teachers developed over a three-year period while working with the Aggie STEM Center. They are proud of these lessons and wanted to share them with you. Second, think of yourself as a chef, a "Teaching Chef," who prepares the ingredients into a bounty for your students to consume. As you read this book, think of it as a cookbook, a "Lesson Cookbook," filled with tempting ingredients whipped into scrumptious morsels. Finally, think of the lessons contained in this Lesson Cookbook as recipes that worked for the chef who lovingly and carefully crafted every aspect of each meal. You are the chef now, so just as you would modify and enhance a great recipe for your own kitchen, feel free to customize these lessons to your classroom and make improvements that you think will be gobbled up hungrily by your students. Then feel free to share your new or improved STEM PBL lessons with the Aggie STEM family at aggiestem@tamu.edu or http://www.aggiestem.tamu. edu. In the future, your lessons could be included on our website or in the next version of this book.

We dedicate this book to our loving families who share us with our avocation, teaching; all the teachers and administrators who have and will work with the Aggie STEM Center; and all those who came before us, on whose shoulders we stand:

Linda, Patra, Charles, Amira, Coco, Pumpkin, Veronica, Sarah, Corinna, Jasper, Patti, Wilma, Beatrice, Michael, Carolyn, Marie, Becky, Susan, Mary, Kristin, Hector, Mary Ann, Dennie, Deann, and Suzanne

TABLE OF CONTENTS

DEBORAH FLOYD

INTRODUCTION

An Inside Look at PBL

WELL-DEFINED OUTCOME

Mission Statement. The purpose of the Emmett J. Conrad Science, Technology, Engineering, and Mathematics (STEM) Academy is to increase the number of students who are college-ready. Conrad High School and it faculty are committed to Project-Based Learning (PBL) through rigorous, relevant, and innovative instruction, enabling students to pursue higher education.

INTRODUCTION/BACKGROUND

I hear and I forget.
I see and I remember.
I do and I understand.
–Confucius

The vision of the STEM Academy at Conrad High School began with the collaboration of Dr. Barbara Ware and Julia Gentry, who worked together to write the original Texas-STEM Grant at the Dallas Independent School District (ISD) central office. These two, along with Conrad's first principal, Donella Perry, planned the structure of the STEM Academy that incorporated the practice of STEM PBL. I am grateful to the pioneering educators whose foresight for the students of the Dallas ISD and the State of Texas made this work possible. Additionally, Mary Fitzgibbon, the principal at the time of writing this chapter, supports these efforts.

I was hired as the STEM Academy Coordinator in January 2007. I came from a science and engineering magnet where I served in many leadership positions such as Math Department Chairperson, School Coordinator, Campus Instructional Leadership Team member, and Teacher of English and Mathematics. I started as an urban high school teacher in 1994.

My own secondary school background contributed to my foundation and provided me with role models. At my high school, the educators were a group of innovators, exploring non-traditional ways to deliver instruction. Students did not have worksheets at my high school. Schedules were innovative as were assignments.

THE CHALLENGES OF LEADING A GROUP OF TEACHERS TO STEM PBL

Conrad High School opened in the fall of 2006 with approximately 600 students. The school served grades nine and ten. The demographics included approximately 96% minority students and over 90% classified as low socio-economic status with a mobility index of 36%. Although these statistics may be reflective of many Texas schools, the school's population differed in the number of immigrants (50%) from around the world, and the school, at the time of its opening, had no peer in Texas public education. It was in this context the STEM Academy would recruit students and teachers during the 2007-2008 year. During this year, one Career and Technology Education (CATE) teacher was in place and two were to be hired for the following year. The core STEM teachers volunteered from the existing faculty to teach in the STEM Academy.

The faculty of the Academy knew one part of the mission was to integrate STEM PBL throughout the curriculum and across all subjects: science, technology, engineering, and mathematics, but the staff expected more and included English and social studies. This group of educators had a good understanding of projects but soon learned STEM PBL did not have a single, clear definition. It was also not clear how STEM PBL would work with the mandatory state curriculum and Dallas ISD's curriculum planning guide. Because it was impossible to set aside the state curriculum or Dallas ISD's curriculum planning guide, the Academy's staff had to integrate and create synergy among the three sets of expectations.

The first attempt to create an integrated project was a crime scene investigation, entitled "V-Day Mystery,". This murder mystery involved identical twins, DNA, blood spatters, forensic odentology, and forensic entomology. This work involved staff from the Texas A&M Aggie STEM Center, Dallas ISD Central Staff, the Dallas Police Department, and the Academy faculty and administration. The project was creative, interesting, and detailed, and it contained the components of an engaging and applicable project. After several meetings and attempts at creating a viable Crime Scene project, it became apparent that a project involving all of the content areas should be postponed until the teachers had first developed an operational understanding of PBL for each of their respective content areas. The initial process was like trying to write a complex symphony before learning to write a simple melody.

During the summer of 2007, after stepping back and reviewing progress, the team decided it omitted a few foundational and preparatory activities. The entire staff had not yet been hired at the time of the original staff's first attempt at writing a PBL; therefore, the staff did not have important input from all content areas. As a result of this review, I determined that we needed to lay a foundation before writing a PBL of any length, depth, or scope. In preparation, I wrote a three-day professional development curriculum with the goal of establishing a series of steps for developing PBL activities and templates for future PBL activities. The Academy faculty participated in this professional development for the purpose of developing a mission statement as an introductory step to focus the future work of the Academy.

Once the mission was established, the faculty participated in team-building activities, reviewed resources, studied Texas public school curriculum, analyzed student data, and identified the basic required content elements of STEM PBL. Establishing a common mission verifying each stakeholder's commitment to STEM PBL was an essential step in showing PBL as a focus of curriculum delivery; STEM PBL work became the standard for lesson-plan development.

The Academy staff also developed team norms to direct behavior in team collaboration. These norms included being on time, respecting diverse opinions, encouraging openness, and participation. Participation in team-building activities helped the staff better understand one another's work habits, strengths, and motivating factors for behavior. Next, the team developed a list of curriculum and PBL resources relating to STEM PBL as a source to build on throughout the year. But STEM PBL had yet to be defined.

THE TEACHERS DEVELOPED QUESTIONS...

What is PBL? How long does a PBL activity last? Where is it done? How many people does it involve? Does it include worksheets? How much does it cost? What is the finished product? Why do a PBL activity? What will the administrators think? Should teachers remain tethered to Dallas ISD's curriculum planning guide? Has this been done before?

We began to answer these questions as we progressed on our journey of discovery through our collaborative work with the Aggie STEM Center. However, there were still important considerations. First, student achievement data drives everything the Academy does. It was only natural that the PBL activities have their foundation in meeting student needs. The staff started with a review of current student data. Teachers identified areas of student weakness in their content areas, which could use the extra reinforcement of concepts STEM PBL would provide. Then the group met to discuss commonalities and how each content area could support the others. Although this meeting raised a greater awareness of what each person had to teach, it did not resolve all the issues. The teachers agreed on planning template components for each STEM PBL; however, the criteria were flexible and others could be added as needed. It was understood the template was a work in progress. The PBL template included the driving question, length of time, depth and breadth of the project, potential connections, and resources to be used.

The product of the three-day workshop was a working collaboration template for future PBL activities. Teachers were given the assignment to develop their first STEM PBL for their discipline. This approach was successful in that the teachers experienced success in the development of a simple project and began to develop insights into how the STEM PBL planning template could be transferred to creating projects incorporating other subject areas.

At the culmination of the three-day workshop and even after three years of implementation one question continues to require attention. The one lingering question that remains is "How does STEM PBL fit into an established, required curriculum where teachers are held accountable for curriculum taught in a specific sequence?" The staff considered the process of coordinating with the established curriculum, the opportunity to think outside the box, and coordinating with the Dallas ISD calendar. However, one aspect that continues to help us is meeting and working with the Aggie STEM Center for technical assistance and support. In fact, one form of technical assistance that was practically important in dealing with matching STEM PBL with district initiatives was teachers' online work with Aggie STEM Center graduate students. This innovative form of support was effective because it fostered communication that matched just in time assistance with teachers' immediate needs. The teachers were able to work with doctoral students in core STEM areas on specific aspects of projects as necessary.

THE BENEFITS OF PBL

The rewards experienced as a result of this collaboration were varied and numerous. The teachers articulated clear expectations, milestones and learning goals were set and met, and the group effort for PBL led to less student behavior problems and more time for learning. Students showed more self-confidence and pride in their achievements and increased participation. This led to increased time on rigorous learning tasks. Their grades improved and all indicators pointed to increased motivation. This was demonstrated in test scores higher than other students' in the district. These successes led to independent thinkers and problem solvers who were college ready. Students have also worked cooperatively and individually on innovative educational activities and real-world products that have helped them make connections across content areas.

THE REWARDS

The rewards experienced as a result of this collaboration were varied and numerous. The teachers articulated clear expectations, milestones and learning goals were set and met, and the group effort for PBL led to less student behavior problems and more time for learning. Students showed more self-confidence and pride in their achievements and increased participation. This led to increased time on rigorous learning tasks. Their grades improved and all indicators pointed to increased motivation. This was demonstrated in test scores higher than other students' in the district. These successes led to independent thinkers and problem solvers who were college ready. Students have also worked cooperatively and individually on innovative educational activities and real-world products that have helped them make connections across content areas.

RESPONSIBILITIES IN MAKING PBL WORK

The main responsibility of the administrator to make STEM PBL work is to become acquainted with the components and how it differs from traditional instruction. These leaders need to be aware of the current trends in high school reform efforts. They must allow additional interpretations and delivery of the established curriculum. Administrators need to understand that teachers will not be in lock-step progression, working on the same worksheet at the same time every day. Administrators should also realize that teachers need some autonomy and control over their curriculum. Overseeing STEM PBL implementation necessitates stepping outside the box and being willing to model instructional delivery that carries the most effective and sustaining impact—learning by doing. This innovative, risk-taking approach yields dividends in participation, improved attendance, fewer discipline issues, increased concept understanding, and improved test scores. Needless to say, teachers who are happy with what they are teaching and how they are teaching are more productive; they help to build the kind of school climate that fosters these same feelings of happiness in students. It is incumbent on the administrator to provide support and protection for teachers as they take on the responsibility of implementing STEM PBL. Administrators have to be willing to be supportive and strong; change is difficult, but student achievement makes believers. An administrator should be gently persistent, create a definitive set of expectations, and make incremental and iterative progress towards the goals. Administrators should model thinking outside the box, relinquishing ownership over private teaching methods, increasing communication, and engaging colleagues in difficult conversations about successfully improving student learning.

PARTNERS AND MOTIVATION

Industry partnerships and their support of PBL made a big difference on academy success. Students became involved in partnerships with a group of engineering companies, the Dallas Fort Worth Semiconductor and Technology Executive and Council. Inspired by student internships sponsored for Academy students, the students worked with engineers and college engineering students on engineering projects. Students were engaged and enthusiastic about coming to school, showed an increase in their knowledge of engineering careers, and developed a greater propensity to view themselves as pursuing an engineering career. One student replied to the question, "What do you know now that you didn't know before the engineering internship?" by saying, "I learned I am more than I thought I was."

During a two-week session, engineers and university engineering students came to the Academy for four days a week to work with students. On the fifth day, students toured engineering job sites. The Academy was the pilot school for this project, and now multiple STEM academies in the Dallas Metroplex are involving their students in this unique opportunity. Through these experiences, projects were developed. Due to the two-week time limitation, many of the projects completed by the students during these internships were short—lasting two to three hours each. These projects contained all the STEM components for a successful PBL activity.

WHERE IS THE STEM ACADEMY NOW?

STEM Academy is in its second year, student test scores have increased both years, and this year the scores are projected to be above district averages. Students beam with pride as they display and explain their projects. They have presented their projects to groups of teachers, professors, and industry professionals. The Academy has produced students who are self-confident, motivated, and college ready—awaiting the next challenge given to them. The administrators and faculty have made a difference in students' lives and in their future.

Deborah Floyd
Coordinator STEM Academy,
Emmett J. Conrad High School, Dallas ISD

NATURE

DAVID BOYD AND MARY MARGARET CAPRARO

1. WHERE DO EARTHQUAKES OCCUR?

WELL-DEFINED OUTCOME

Students will conclude that there is no difference between locating the x and y coordinates when graphing the slope of a line in algebra and finding locations of earthquakes (EQs) worldwide using latitude and longitude coordinates. Students will discover the three-dimensional image of the depth of the EQ when coupled with time.

INTRODUCTION/BACKGROUND

This introductory activity assists with the understanding of positive and negative latitudes and longitudes similar to graphing x and y in algebra. The location of a seismic event requires latitude (north and south) and longitude (east and west) and depth below the surface of the earth. The fourth dimension is time. When did the event occur? These data can be "coordinated" to facilitate students' interdisciplinary discovery of the nature of EQs through science, language arts, and mathematics. Students will cover a world map using a clear plastic drop cloth and self-adhesive stickers of various shapes and sizes to indicate the magnitude of the EQs they locate on a map. The project will focus on researching EQs of magnitude four and above from the previous month. Students are often not comfortable with locating places on the earth using latitude and longitude. Currently, the process of instructing students about spatially locating themselves is considerably easier than before Google. Before students are allowed to access Google Earth, they will be asked to locate the epicenters of EQs on the United *States Geological Survey (USGS)* website: *http://earthquake.usgs.gov/eqcenter/recenteq sww/Quakes/quakesall.php* and the Incorporated Research Institutions for Seismology (IRIS) website: http:// www.iris.edu/dms/seismon.htm. As EQs are detected worldwide, the date, time (Coordinated Universal Time [UTC] – formerly Greenwich Mean Time [GMT]), epicenter, depth, and map location are indicated by the USGS and IRIS. By combining these data in Microsoft Excel, Students can easily identify the latitude and longitude and determine the magnitude for each EQ using the data on the spreadsheet. Each different magnitude for the EQs will be determined by the students from the USGS list. Students will indicate the magnitudes by using stickers on the drop cloth overlay. The stickers identify the magnitude, location, date, time, and depth of the seismic event. Students will be expected to design a legend with this information. Finally, after the students have collected the data, organized it into data sets, made inferences from the data, and communicated their discoveries to their fellow students, the project will digress into its various adaptations and extensions. Students could also create pie/bar graphs showing which earthquakes had the highest magnitudes, if the locations were in the southern or northern hemispheres, and lastly they could graph out dates on a scatter plot.

OBJECTIVES

Social Science

- The student uses geographic tools to collect, analyze, and interpret data.
- The student creates thematic maps, graphs, charts, models, and databases representing various aspects of the United States and the world.
- The student poses and answers questions about geographic distributions and patterns shown on maps, graphs, charts, models, and databases.
- The student understands how physical and human characteristics of place can change.

Mathematics

- The student gathers and records data or uses data sets to determine functional relationships between quantities.
- The student represents relationships among quantities using concrete models, graphs, and diagrams.

R. M. Capraro, M. M. Capraro, J. Morgan and J. Scheurich, (eds.), A Companion to Interdisciplinary STEM Project-Based Learning: For Teachers by Teachers, 03–10.

- The student interprets and makes inferences from functional relationships.
- The student identifies general forms of linear ($y = x$) parent functions.
- The student uses spatial reasoning to express generalizations about space and spatial relationships.
- The student uses properties and relationships, location, direction, and orientation of figures by finding areas of regular polygons, circles, and composite figures.
- The student analyzes geometric relationships in order to make and verify conjectures.

Language Arts

- The student uses prewriting strategies to generate ideas, develop voice, and plan.
- The student compiles written ideas and representations into reports, summaries, or other formats and draws conclusions.

GIS Elective/Tech Applications

(Alternatively, EQ data can be placed on an Excel worksheet and imported directly into ESRI Arcmap. The same project being performed with the drop cloth EQ map can be done electronically).

- The student assesses how people's changing perceptions of geographic features have led to changes in human societies.
- The student understands how to use geographic knowledge, skills, and perspectives to analyze problems and make decisions.
- The student understands how spatial features influence human behavior.
- The student knows how to use technology to represent and interpret Earth's physical and human systems.

MATERIALS NEEDED

Drop cloth: thin, clear enough to see through

- Computer (for locating the list of EQs)
- Printer
- Internet access
- Journals
- Stickers of various kinds, sizes, and shapes
- A large map table or magnets to hang the map on a magnetic white dry-erase board
- Map of the world*
  *See information about ordering the map in the appendix

Materials Needed for Extension Activities

- Blue painters' tape
- Blue recycling polyethylene bags
- Sticky mailing labels
- Dry erase markers

Vocabulary. Earthquake, magnitude, subduction zone, Modified Mercali Magnitude Scale, Richter Scale, plate tectonics, convergent, divergent, transform, seismograph, seismic wave, *P* wave, *S* wave, surface (Love) wave, tsunami, trench, volcano, mountain range Mt. Pinatubo, Mt. St. Helens, Mount Ranier, Fujisamasan, Alfred Wegener.

The concept of x and y coordinates is very clear to show on a standard geographical map of the world, which nearly every Social Science classroom in the U.S. contains. This type of map rarely needs a thin plastic drop cloth to protect it; dry erase markers of varying colors can achieve the same purpose and be easily cleaned away. (Test an inconspicuous spot on the map before using markers to ensure that the ink will come off). Make sure the map is secure and not hanging from the ceiling. Students will need to stand on something in order to place the stickers. Follow obvious safety precautions at all times.

Time frame for the project. This project takes place over five days, including the extension activities.

DAY 1

Engagement

The teacher starts by saying, "Last night about a dozen EQs occurred. Have you ever experienced an EQ?" and may then pull up a video of an EQ on TeacherTube or NOVA. The teacher begins the project by having an open discussion about how the environment shapes the way people see the world. A dialogue with students to generate ideas and thoughts should be encouraged. Next teachers should have students discuss where they anticipate finding multiple EQs using a variety of maps to locate places where they suspect EQs occur. The use of maps from textbooks can be helpful in assisting students to preliminarily locate EQ activity. Students should make predictions about where tomorrow's EQs will occur. Predictions can be recorded in student journals.

DAY 2

Exploration

Next, students will research and create a list of the past 30 days of EQs with magnitudes of four and above. Students will list these active EQs in their journals. Each entry will contain the date and the latitude and longitude coordinates of the EQ activity. A physical location should be included as well—such as Sicily or the northern Atlantic Ocean. An example of the journal format is shown in Figure 1. The students may organize their data in any way that makes sense to them. This model is provided as just one example. To add more rigor, have students switch journals and have the groups make pie charts or bar graphs on location and depth) Also, students could create scatter plots on dates using another groups journal.

Journal

Date	Latitude	Longitude	Location/depth

Figure 1. Possible journal entry model for EQ activity.

Journal assessment. The teacher or student peers should evaluate the journal entries using the Journal Checklist in Figure 2. Points will be awarded for each item included. Thus, a perfect score would be 100 (30 points for dates, 30 for latitudes and longitudes, 30 for locations, and 10 for organizational setup of the journal).

Dates are listed	/30
Latitude and longitude are accurately listed	/30
Locations are included	/30
Accurate journal setup	/10
Total	/ 100

Figure 2. Journal checklist.

DAY 3

Explanation

On this day, the teacher will introduce the USGS hazard maps (http://earthquake.usgs.gov/) and the IRIS EQ activity map (http://www.iris.edu/hq/) to identify locations of seismic activity. Students should then draw conclusions about patterns and discuss those conclusions by comparing and contrasting seismic events and where they occur. Some questions to ask to stimulate discussion are: Where do the deepest EQs occur and why? Where do the shallowest EQs occur and why? Do all EQs occur on land? Why? Are EQs evenly dispersed throughout the world or are they concentrated in some active EQ zones? Students should use their journals to look at the similarities of the geographic areas where the EQs were found.

Additionally, ask the students to find three earthquakes that make a right triangle and measure all the sides to prove Pythagorean Theorem. During science discuss the different layers of the earth in which earthquakes occur. In what layers of the earth do they start? Why do you all think this happens? Is there any human activity that could increase the occurrence of earthquakes? (Thus natural gas drilling debate in Cleburne). In the engineering curriculum, there is a lesson related to creating a structure and then putting it on a shake table to see if it fails or not. This could be part of this project to increase rigor, students would actually act out an earthquake.

Student activity for Day 3. Students can be assigned one of the questions above to research in particular, focusing on patterns. For students' next journal entry, teachers can assign topics such as depth (i.e., EQs deeper than 100 m, EQs shallower than 50 m), population centers (i.e., cities, towns), or location (i.e., trenches, land, water, mountains, near volcanoes). Students can select one specific category. They will need to find five EQs that fit this specific category and write a coherent paragraph describing their observations of the pattern. Students will evaluate each other by exchanging journals and assessing their peers using the following criteria: are five EQs listed, do all fit the pattern, and does the paragraph address the patterns, which should be consistent with the criteria? Time can be provided for students to justify their pattern findings with their peers. If time allows, this can be turned into a class discussion.

DAY 4

A table-sized map for each group can be used to help students locate and understand the latest EQ activities. Each group of four students will work with a map centered on the Pacific Ocean. Groups of four can work cooperatively on this project. For one possible organizational structure, divide students in the group by the EQ magnitudes of four, five, six, and seven. For more data points, include EQs from 3.0 to 3.9. There are only a few magnitude eight EQs annually; thus, groups of four can study the majority of EQs between magnitudes four to seven. As students are working in their groups, the teacher will circulate to determine their participation and cooperation in researching EQs using the eight-point participation rubric in Appendix A.

DAY 5

Evaluation

Student groups (the four, five, six, and seven magnitudes) can take turns finding the locations and EQs on the class's large drop cloth map. Students will come up in groups to the board and place their colored stickers on the correct latitudes and longitudes. A key or legend will display the magnitudes by color. Depths will be written on the stickers in meters/miles. Ultimately, the object will be to compare the drop cloth map and the U.S. geological survey map as to the general pattern of EQ activity. At this time, results should be discussed with students so that they will be able to compare and contrast trends in EQ locations, population centers, and depth. The 16-point presentation rubric (see Appendix B) will be used to evaluate student group performance.

Extensions

A variety of extensions can accompany this project-based learning activity; extensions of the project are as numerous as the coordinates themselves.

Students may write descriptions of the patterns they discovered in their study of EQ zones. How are EQs similar/different in Chile compared to California? How do the depths of average EQs compare/contrast in different geographical areas of EQ activity?

Use blue painters' tape to mark off the floor of your classroom with the Equator and Prime Meridian. Use brightly colored mailing labels to mark latitudes and longitudes on top of the blue tape in increments of 30 degrees for meridians and parallels. Have students "adopt" an EQ and go stand on its coordinates on the map on the floor. Have all students participate and find out where the crowds of EQs are.

Have students put a blue plastic bag over a standard-sized classroom globe and follow the same procedure as with the drop-cloth map, but find the locations of EQs on the globe. When finished, students should cut the globe bag down on the seams and compare/contrast with the flat map results. Next students can write their observations.

Ask students to respond to a focus question for expanded writing, such as: What kind of structure would best withstand the average EQ? Describe a rail or transportation system that could withstand EQ shaking. How could communities better serve themselves in the event that an EQ took place? How would you design your schools, hospitals, or essential services if you were the mayor of a city prone to EQs?

Google Earth can be used to "pinpoint" or mark locations where EQs occur, but Google has that feature as well; use Google to plot population centers and approximate numbers of people affected should an EQ of a catastrophic magnitude occur near major cities like Tokyo, Santiago, Istanbul, San Jose, Manila, and Jakarta.

Students can make a graph of EQ depths from a "side view" to show the plunge of the Pacific plate beneath various locations along the Ring of Fire. How do the sharpness of the angles of decline (slope) of the subduction zones compare to each other? How would you represent the angle of slope in a formula for each subduction zone?

Students can calculate cities with populations over 300,000 that are in high EQ zones. By combining these populations within affected areas, students can calculate the total number of people who are impacted by EQs regularly.

Students can map out the earthquakes as a point feature and symbolize them using magnitude if GIS Arcmap or free Arcviewer is available. Therefore, the larger the dot the higher the magnitude.

Finally, teachers can ask students to use vocabulary words to make haikus and diamantes.

APPENDIX A

Participation Rubric: 8 Points

	Seismologist 4	*Student Seismology 3*	*World traveler 2*	*Geology Newby 1*
Participation	Student was actively engaged during the project.	Student was engaged in most of the project.	Student had to be told a few times to get back on task.	Student was off-task most of the project.
Cooperation	Student proactively contributed to the group by offering EQ sites.	Student contributed to group by offering EQ sites.	Student rarely contributed to the group by offering EQ sites.	Student never contributed to group by offering EQ sites.

APPENDIX B

Presentation Rubric: 16 Points

	Seismologist 4	*Student Seismology 3*	*World traveler 2*	*Geology Newby 1*
Uses visuals clearly and effectively	Highly developed, memorable visuals, illustrates main point. Manages visuals well.	Visuals clarify and illustrate main point. Presentation is memorable and visuals are neat and well organized.	Visuals clarify and illustrate main points.	Visuals not tightly linked to presentation; do not support or clarify main points.
Communicates and stresses main points	Communication is articulate and well thought out. Main points are very clear.	Communication is good; most of the main points are covered.	Communication is somewhat lacking, but main points are still evident.	Communication is poor and main points do not appear to be fully covered.
Graphics - Relevance	All graphics are related to the topic and make it easier to understand.	All graphics are related to the topic, and most make it easier to understand.	Not all graphics are related but are easy to understand.	Graphics are hard to understand and not related to the topic.
Content Knowledge	Locations (10) are accurate and consistent; geographic landforms are identified.	Locations (8) are fairly accurate and consistent; geographic landforms are identified.	Locations (6) are accurate and fairly consistent; geographic landforms are identified.	Locations (<4) are inaccurate and non-consistent; geographic landforms are identified.

***Map information:** The World - Copyright U.S. Government 1998. Map number 33992 is available through the USGS toll free: (888) 275-8747. The maps can be ordered by fax: (303) 202-4693.

The feature of this map is that it is closely centered on the Pacific Ocean. The concept of quadrants less graphic with this map requires the dependence on understanding positive and negative latitudes and longitudes that are NOT centered on the Equator and Prime Meridian. The challenge is to have students approach exploring the location of EQs from a different point of view. If this type of map is unavailable, teachers may prefer to use a standard map of the world. Simultaneous plotting can take place with “AVID quick writes,” identifying differences and similarities.

Students locating EQs on the class drop cloth maps

David Boyd
World Geography Teacher,
Emmett J. Conrad High School, Dallas ISD

Mary Margaret Capraro
Teaching, Learning, and Culture,
Texas A&M University

ROBERTO GARCIA AND ALMA F. RANGEL-CHAVEZ

2. MODELING AND ESTIMATING RAINFALL

Why are rainfall data and analysis so important in predicting weather patterns?

INTRODUCTION

In this increasingly complex world, people have to make decisions based on huge amounts of data. To obtain better outcomes from the analysis of this data, scientists build mathematical models. By modelling real problems, scientists can predict and get a deeper understanding of events affecting human life (For example, the use of mathematical models for weather forecasting).

Weather impacts many human activities such as agriculture, urban planning, economy, recreation, etc. One of the most important factors within the weather cycle is precipitation or rainfall. This highly variable factor is related to the amount of fresh water available on the planet. Although water is essential to all living organisms, precipitation can yield catastrophes that can harm or even annihilate entire societies. For these reasons, it is important to understand how precipitation is measured and how to analyze precipitation data to obtain predictive models.

In this project, students will first explore the implications of rainfall on people's lives, the reasons for measuring rainfall, and the tools used to collect precipitation data. After they have selected the place/region and time period they want to work in, they will select a viable database for their own questions. They will learn how to download, manage, and analyze their selected data to create a viable prediction of rainfall.

Students will learn that although it is difficult to predict rainfall to an exact science, they may be able to make an intelligent estimation of rainfall patterns using statistical analysis (e.g., mean, median, mode, standard deviation, maximum, minimum, and correlations). With the use of Excel's least square algorithms, they will fit functions to their data. For instance, students can use linear, quadratic, n^{th}-order polynomial, exponential, or piecewise functions to fit their data.

The teacher also can promote the use of TI Navigator software and hardware to send data back and forth to the teacher for inspection. The teacher will then send the data back to the student to construct maps and graphs (scatter plots, bar graphs, histograms) using software already in existence.

After students have found the most appropriate model for their data, they will have to defend and present their findings in an oral presentation. They will need to justify their analysis and make conclusions addressing the potential impact of rainfall on society and the accuracy of their model.

OBJECTIVES

Mathematics

- Select the appropriate measure of central tendency to describe numerical data and justify the choice for a particular situation.
- Collect and organize data, make and interpret scatter plots, fit the graph of a function to the data, and interpret the results.
- Use regression methods available through technology to describe various linear, quadratic, exponential models; select the most appropriate model, and use the model to interpret information.
- Identify changes in weather patterns based on analyzed data.
- Recognize misuses of numerical information.
- Use interdisciplinary connections, both within and outside mathematics, multiple representations, applications and modeling, reasoning, and computation in problem-solving contexts.

R. M. Capraro, M. M. Capraro, J. Morgan and J. Scheurich, (eds.), A Companion to Interdisciplinary STEM Project-Based Learning: For Teachers by Teachers, 11–26.

Earth and Space Science

- Demonstrate scientific reasoning and logic by analyzing how science explains and predicts the interaction and dynamics of complex earth systems.
- Investigate and understand the energy transfer between the sun and the earth and how the earth's atmosphere drives weather and climate on earth. Key concepts include observations and collection of rainfall data.

Social Science

- Compare ways humans depend on, adapt to, and modify the physical environment using local, state, national, and international human activities in a variety of cultural and technological contexts.
- Improve geographic literacy and become more informed global citizens.

Scientific Processes

- Plan and implement investigative procedures including asking questions, formulating testable hypotheses, and selecting and using equipment and technology.
- Communicate valid conclusions; construct graphs, tables, maps, and charts using tools including computers to organize, examine, and evaluate data.

Language Arts

- Organize ideas in writing to ensure coherence, logical progression, and support for ideas.
- Use prewriting strategies to generate ideas, develop voice, and plan.
- Compile written ideas and representations into reports, summaries, or other formats and draw conclusions.

MATERIALS NEEDED

- Personal journal
- Computer with Internet access
- Graphing utility software (i.e., Microsoft Excel worksheets or calculators like TI Calculators or the TI Navigator)
- The statistical glossary provided by HyperStat Online (http://davidmlane.com/ hyperstat/glossary.html)
- The weather glossary provided by the Weather Channel (http://www.weather. com/glossary/)

STUDENT INTRODUCTION

The Aguirre family moved from the San Francisco Bay Area of California, to Dallas, Texas. The Aguirre family wanted to move because they had just lost their home due to a flood and also the children wanted to live close to their relatives. When they arrived in Dallas, they decided first to move in with relatives while the father looked for a new home. The flood they experienced in San Francisco taught them not to make the mistake of buying a home in a low area, where there is a high risk of flooding. Adela, the eldest child of the Aguirre family, enrolled in the school her cousin Lucy attends. In their mathematics class, they are learning about descriptive statistics and functions, and the teacher is implementing a project to investigate rainfall in the Dallas area. Students, working in groups of two or three, will conduct a regional rainfall analysis of a Dallas area (or surrounding) of their choosing. Then, they will compare different locations in terms of the amount of rainfall the locations get and decide on the safest place to avoid flooding in Dallas. Adela and Lucy will work as a team for this project, using recent rainfall data to explore and understand the rainfall patterns of their favorite place in Dallas. This project will help Adela's father as he searches for a location for their home where the flood risk is minimum.

ENGAGEMENT

To hook students' interest in this project, teachers need to find possible learning experiences to help students see the relationship between the project's objective and their own lives, interests, curiosities and prior knowledge. One way to do this is by providing the class with photos. The first photo shows how heavy rain can cause flooding. The second photo shows a drought (see Figures 1 and 2).

Figure 1. Flooding.

Figure 2. Drought.

The teacher starts off the project by having an open discussion about the pictures. The teacher can ask questions like, "What do you notice here? What questions can you pose to better interpret the images displayed in the two photos of heavy rain and drought?" Students can conjecture on one or more variables related to the photos. For example: is flooding related just to temperature? What about the relationship with the altitude or geographic location? Later, the teacher should indicate that students should answer the following questions individually: What is precipitation? When, where, and why does it happen? Is there any difference between precipitation and rainfall?

Students will have the opportunity to record their own answers in a journal before sharing them with the class. The students' journals will serve as a tool to assess students' reflection on their own learning process. Also, it will allow teachers to learn what students already know: their accurate conceptions as well as their misconceptions.

While students discuss their ideas, the teacher can write them on an overhead or a projector. The teacher asks students what the consequences might be of heavy or very light rainfall and the relationship of rainfall to flooding. After discussing the factors affecting flooding, the teacher might mention that in an effort to simplify the problem, they will assume that only rainfall affects the amount of flooding.

Another important topic to discuss is how rainfall is measured. For example, the types of technology used or the measurement scale used for determining the amount of rainfall can be discussed. The measurement complexity of precipitation can also be addressed at this time. Then, the teacher should create a dialogue to develop student interest in learning how much rain fell around their homes during previous years.

Further supporting material and student engagement activities can be obtained by accessing free Internet resources (see Table 1). From this list, teachers can select the appropriate videos or lessons to reinforce the misconceptions or cognitive deficiencies students manifest in their journals or during class discussion.

Table 1. Weather Internet sources

Title	*Content*	*URL*
Background and History of Rain Measurement	This handout provides the history of rain measurement. This reading will enable students to discuss the major events, people, and technology involved in measuring rainfall in the past as well as possible new technology for the future.	http://www.cocorahs.org/Media/docs/THE_HISTORY_OF_RAIN_MEASUREMENT.doc
Global Precipitation Measurement	This website study global precipitation through satellite-based data. A video resource called *The Promise of Rain* shows students their relationship to precipitation and how it affects their lives. Interesting information like the science of measuring precipitation, why it matters, and the advantages of using a space-based remote sensing instrument to obtain precipitation measurements are available on the main page of this website.	http://gpm.gsfc.nasa.gov/images/gallery/promiseofrain.mpg
Miami Museum of Science	*Making a Weather Station* is a site provided by the Miami Museum of Science. Even though this site appeals mainly to children, older students can benefit from the clear and concise information on variables that cause changes in the weather.	http://www.miamisci.org/hurricane/weatherstation.html
Precipitation Estimates, Part 1: Measurement	Through the use of rich illustrations, animations, and interactions, this module provides an overview of the science of precipitation estimation using various measuring platforms. Students learn the strengths and limitations of technology for remote sensing of quantitative precipitation estimation (QPE) like radar, satellite, and rain gauges. (Preregistration is needed in order to access this website.)	http://www.meted.ucar.edu/hydro/precip_est/part1_measurement/
Weather and the Built Environment	This course provides an overview of the evolution of the modern urban environment with a focus on its impact on the urban watershed, air quality, and climate. Specifically, *Unit 2: Impact on the Watershed* will help students explore how the built environment affects the water that moves through an urban watershed. (Preregistration is needed in order to access this website.)	http://www.meted.ucar.edu/broadcastmet/wxbuiltenv/

EXPLORATION

The teacher presents the Aguirre family's story to students and asks them to assume they are in Adela and Lucy's class. Students are asked to get in small groups of two or three and brainstorm about collecting data, producing graphs, and data analysis. They should write down anything they think they need in order to collect and produce these graphs. On the overhead projector, the teacher can list some of the items that students put down about creating their projects.

After that, the teacher can have students explore the available data about rainfall on the Web. The teacher might make some suggestions to the students about where to start looking for the data on the Internet. Some of the websites that provide average numerical precipitation data are: Country Studies (http://countrystudies.us/united-states/weather/) and WorldClimate (http://www.worldclimate.com/).

The aforementioned websites are good starting points because they contain easy to search and easy to analyze monthly rainfall data. However, they just provide information for major cities or counties and the range of time for which the data is averaged is fixed. Although students may not be interested in the periods of time or locations available through these sources, these databases can engage students and help them identify questions worth pursuing, researching, analyzing, and reporting in their findings. Figure 3 (a) shows Dallas' rainfall data from the website WorldClimate, averaged from 1961 to 1990, while Figure 3 (b) shows CountryStudies' data for the same location averaged from 1986 to 1998.

The exploration and inquiry process of the data shown in Figure 3 can start with teacher induced questions such as: "Why do we use length units to measure rainfall? What are the instruments used to obtain this data? If a city smaller than Dallas reports the same average rainfall in inches, does this mean that both cities received the same amount of water from the rain? If you want to know the average volume of water falling in Dallas what other information do you need? Why didn't the data change significantly between the two sources of information even though the time range used to obtain the averages was different?"

The teacher can also motivate the use of descriptive statistics (mean, mode, median, standard deviation, etc.) and create graphical representations such as scatter plots. This process will enhance students' verbal, written, and graphical mathematic communication skills.

Weather station **DALLAS LOVE FIELD, DALLAS COUNTY** is at about 32.85°N 96.85°W. Height about 134m / 439 feet above sea level.

Average Rainfall

	Jan	Feb	Mar	Apr	May	Jun	Jul	Aug	Sep	Oct	Nov	Dec
mm	44.7	58.9	81.7	98.2	128.0	89.1	60.1	58.4	90.6	97.7	60.1	48.2
inches	1.8	2.3	3.2	3.9	5.0	3.5	2.4	2.3	3.6	3.8	2.4	1.9

Source: DALLAS LOVE FIELD, DALLAS COUNTY data derived from NCDC TD 9641 Clim 81 1961-1990 Normals. 30 years between 1961 and 1990

Map of the area around DALLAS LOVE FIELD, DALLAS COUNTY from tiger.census.gov.
Locations outside the continental US are not mapped.

(a)

Dallas Weather

US Geography / US Weather / Texas Weather / Dallas

	Jan	Feb	Mar	Apr	May	Jun	Jul	Aug	Sep	Oct	Nov	Dec
Avg. High	54°	58°	68°	77°	84°	91°	95°	95°	87°	78°	66°	57°
Avg. Low	34°	38°	47°	56°	64°	71°	76°	75°	68°	57°	46°	37°
Mean	45°	48°	58°	67°	74°	82°	86°	86°	78°	68°	57°	48°
Avg. Precip.	1.8 in	2.3 in	3.3 in	3.9 in	5.0 in	3.5 in	2.4 in	2.3 in	3.6 in	3.9 in	2.4 in	1.9 in

Degrees in Fahrenheit

(b)

Figure 3. Snapshots of rainfall data obtained from websites (a) Worldclimate and (b) CountryStudies.

Another type of rainfall data can be obtained from contour or color maps. Figure 4 shows examples of these maps for rainfall in Texas from the following sources: CoCoRaHS (http://www.cocorahs.org/), MyNasaData (http://mynasadata.larc.nasa. gov/data.html) and National Weather Service (http://water.weather.gov/). Figure 4 (a) shows single day precipitation data from fixed positioned gauges (CoCoRaHS), whereas Figures 4 (b) (MyNasaData) and 4 (c) (National Weather Service) show temporal averages of precipitation data from multi-sensor sources. "Multisensor" indicates the use of combined information from gauge and radar instruments, and in areas where there is limited or no radar coverage, satellite precipitation estimates (SPE) can be incorporated (National Weather Service, 2009).

It should be noticed that although these maps are useful for spatial estimation of precipitation, they lack the rich temporal information available on data sources like the ones in Figure 3. However, data found on the websites in Figure 3 is only measured through time in a fixed point in space.

Maps shown in Figure 4 will help students explore this "new" way of obtaining quantitative information. Possible exploration and engaging questions are: Why is Figure 4 (a) so different from Figures 4 (b) and 4 (c)? What measuring devices were used to obtain these maps? How from multiple, punctual gauge data we can obtain a continuously colored map? Which type of data representation (numeric or map) is better for obtaining temporal weather patterns? How can numeric data (Figure 3) be compared to color maps' information (Figure 4)?

Although it was stated that a single contour map lacks temporal information, some data sources like National Weather Service offer the possibility of choosing the time interval over which the rainfall data is averaged. The effect of using different amounts of time to obtain the average rainfall is shown in Figure 5. These maps, along with the teacher's guiding questions, can help students understand colored map reading. For example, "Why does the color scale change in the different cases of Figure 5? Why do the averages of a short amount of time produce regions with no rain? Does it mean that it never rains in those regions? Which map from Figure 5 would you use to estimate the amount of rainfall for a given region? Why?"

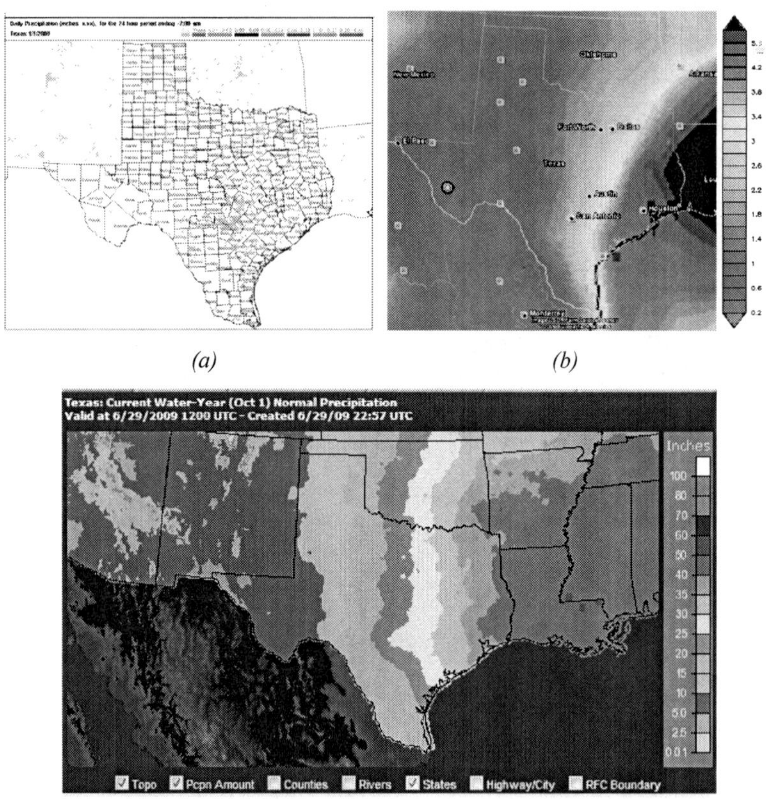

(c)

Figure 4. Snapshots of precipitation maps for Texas obtained from different sources: (a) CoCoRaHS, (b) MyNasaData and GoogleMaps, and (c) National Weather Service.

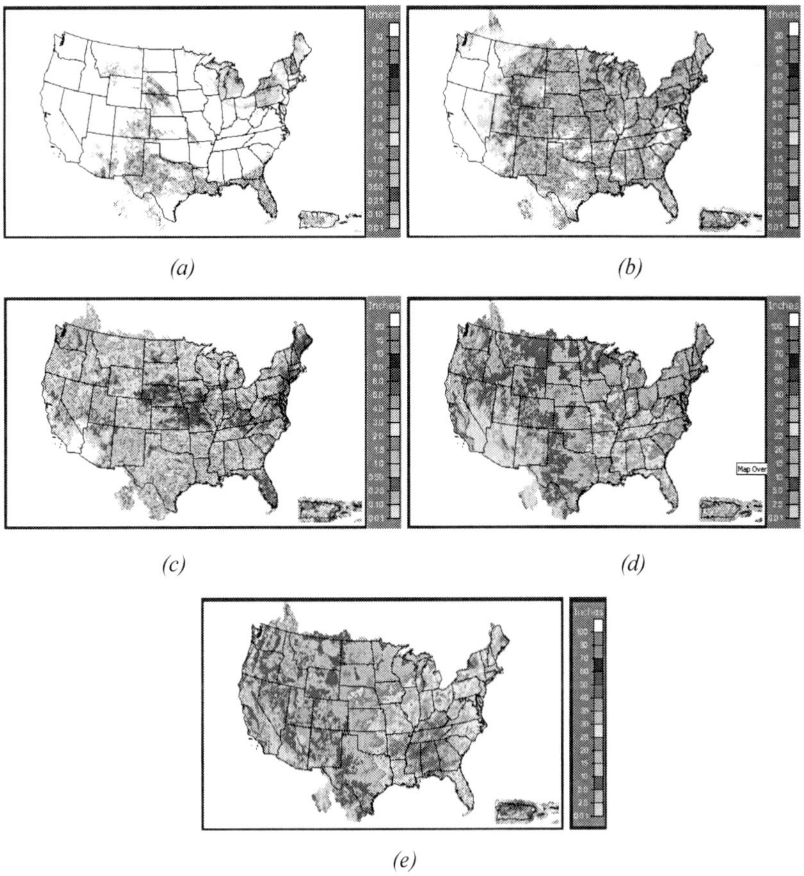

Figure 5. Contour maps of average rainfall in the United States averaged over a time interval of (a) one day, (b) one week, (c) one month, (d) one year, and (e) about 30 years.

Multiple data representations help students critically evaluate data validity, collection, and interpretation. Moreover, students will realize that the meaning of data depends on context and experience. This will lead students to select the best data representation for their specific needs and understanding. Manduca and Mogk (2002) discuss in more detail the benefits of teaching with data.

EXPLANATION

After students gain understanding about the complexity of both rainfall phenomena and data acquisition, they need to venture into statistical analysis and mathematical modeling. The complexity of the research project will depend on students' inquiries and curiosity after the exploration. That is, data source selection will depend on the parameters that students wish to analyze. For example, they could compare monthly, annual, or seasonal mean changes of rainfall by season, etc. Students' inquiries, hypothesis, and objectives must be recorded in their journals for future self-evaluation.

If time restrictions exist, it is preferable to select simple data, such as the sources given in Figure 3. However, this data does not allow much statistical exploration nor does it enhance the data-managing skills needed nowadays to solve real world problems. Therefore, the teacher should recommend that students select a database rich in details and big on content. This will not only improve data handling and data analysis skills, but it will also provide a better understanding of the studied phenomena.

A recommended database can be found at the National Oceanic and Atmospheric Administration (NOAA) website (http://www.noaa.gov/). This website is one of the largest sources of climate information on the country. Finding the data of interest in this much information can be an overwhelming task; therefore, the next paragraphs include step by step instructions for downloading rainfall data.

Step 1 □ Navigate to the weather data

Access the National Climatic Data Center (http://www.ncdc.noaa.gov/), and in the menu area along the left side of the screen under the heading "Purchase," click Most Popular. On the new page that opens, notice in the first paragraph that this data is available for free if you connect to the NOAA website from a server that is in an edu, k-12, gov, or mil domain.

Scroll down to the section labeled Most Requested I. This data collection is labeled Surface Summary of the Day. At the left end of the section, click on the link that reads 1. 3200-3210/CDO. On the next new page that opens, click the button with the text that says, "Continue with SIMPLIFIED Options."

Step 2 □ Choose the State, Weather Station, and Time frame, and Order the Data

Each group of students can select any state, county, weather station, and timeframe according to their interests. In this example, select Texas and Specific County, and then click Continue. Scroll down to the Collin county weather station, select it, and click Continue. Select the date range from 2006 to 2008 and click Continue. This will open a new window where you can review the order. In the same window click the inventory review box, enter your e-mail address and click Submit Request.

Note: The file size will depend on the years selected to download. Selecting a large date range will provide a really big file because daily information is provided.

Step 3 □ Downloading Weather Data

Once you have submitted your request for weather data, a new web page will open confirming your order. It will also have a link to your data. The data may not be available immediately, but it should be available within a matter of minutes. You will receive e-mail notification when the data is available.

When you click on the link to download the weather data, a new screen will open with links to four data sets (you will receive these same four links in the notification e-mail). The top two links (web form and ASCII file) link to the data you need.

Right-click to download the ASCII file data to your computer. Save the ASCII file on your hard drive with a descriptive name. Let's use "Texas-Collin-2006_2008.txt." The ASCII file holds the weather data in a format that can be imported into Excel and graphed.

Step 4 □ Opening and Handling the Rainfall Data

In Excel, select the File menu and then Open. In the window that opens, select the option "Enable all Readable Documents," and open "Texas-Collin-2006_2008.txt." Use the text import wizard to format the data, and save the reformatted file as a Microsoft Excel file with the file name "Texas-Collin-2006_2008.xls."

Step 5 □ Format "Texas-Collin-2006_2008.xls"

The letter *T* in column M stands for Trace and means that there was precipitation on that date, but that it was less than one hundredth of an inch. Replace it with a numerical value. Use the Find and Replace functions of Excel to replace the *T*s in column M with the numerical value 0.

The number 999.99 is a code indicating that no observation was made. You have to replace it with another value (possibly 0).

After following the previous steps, students can explore their downloaded data using Excel or the TI calculator. They will be asked to calculate mean, variance, standard deviation, maximum, minimum, correlation, and trend estimates for rainfall data. Possible student explorations are shown in Figures 6 to 9, where rainfall data for Collin County, Texas, was used. The analyzed information provides daily weather records from 2006 to 2008 on three different weather stations (Frisco, Lavon Dam, and McKinney).

Figure 6 shows the monthly average rainfall for three years (2006, 2007, and 2008). The weather stations' information was averaged to obtain a single value for each month. Although students may hypothesize that the rainfall behavior follows a similar trend every year, different patterns can be observed. After plotting their data as in Figure 6, students should realize the difficulty of predicting and modeling rainfall due to its variability. This data can also help introduce the concept of a statistical outlier. The teacher can point out the abnormally large amount of rain that fell in June 2007 and ask students to investigate the consequences of such a large amount of rainfall (www.cbsnews.com/stories/2007/06/29/national/main2996615.shtml).

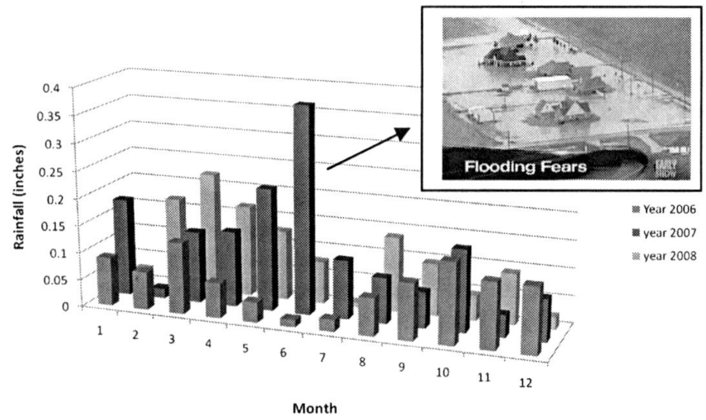

Figure 6. Average monthly rainfall for years 2006, 2007, and 2008 for Collin County, Texas.

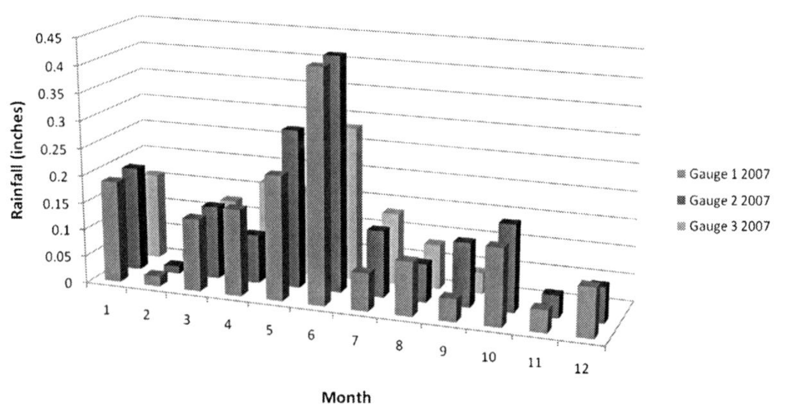

Figure 7. Comparison of the average monthly rainfall of three measurement stations in Collin County, Texas, in 2007.

The teacher can encourage students to question the validity of the outlier; possible teacher questions are: "June 2007 rainfall was almost 50% greater than any other month. Do you trust that value? How can you be confident of that measurement?" Such questions can be answered with the plot in Figure 7. This plot shows an average monthly comparison of three different weather stations (Gauge 1, Gauge 2, and Gauge 3) for the year 2007. For the whole year, the three different weather stations reported a similar amount of rainfall (even for June), with small discrepancies between them. This can serve to justify the validity of the June 2007 measurements.

Further analysis of Figure 7, and the fact that each weather station is in a different location, can lead to conclusions regarding the selection of a region with the least probability of flooding. Students should notice that Gauge 3 reports the lowest amount of rainfall for almost all months, indicating that a region close to Gauge 3 should be selected to have a lower probability of flooding.

Once the students have accurate plots of their data, each group needs to obtain a rainfall mathematical model. The students, then, talk about the shape of the data on their plots. Does the data follow a line, a parabola, a cubic function, or something else? The students should share their graphs with their teacher and discuss the shape of the data.

Students can test their ideas using the fitting algorithms that Excel provides. They can select several functions such as linear, quadratic, exponential, or n^{th}-order polynomial functions to fit their data. Students might be at a stage where they only know how to write linear and quadratic functions. In such a case, the teacher needs to intervene and teach about other kinds of functions that can be used in their models. Figures 8 and 9 show examples of Excel's data fitting capabilities.

Figure 8 shows a column plot of 2006 average monthly rainfall data. It also shows a fourth-order polynomial function fitting the data with R-square (R^2) = 0.799.

Students may recognize different patterns between specific intervals; in that case, a piecewise analysis can be used. Figure 9 shows a piecewise analysis. The data was divided into two sections, and for each section a linear and quadratic function were fit to the data. Students should realize that the R^2 for linear functions are lower than those for quadratic functions. Teacher can explain that a goodness of fit value (R^2) close to one indicates a good fit to the data; however, that value may produce completely unexpected results between data points. This is particularly true for high order polynomials. Therefore, first or second order polynomials are recommended, because they may produce a more realistic model of the input data.

The teacher can promote the use of different data representations using software, calculators, or descriptive statistics. For example, a column plot and a scatter plot of the same information are shown in Figures 7 and 8 respectively.

After students have obtained their functions for the data of interest, they will need to report their findings in a PowerPoint presentation. In their oral presentation, students need to not only critically evaluate the pattern and validity of their data and their chosen model, but also address the importance and potential impacts of rainfall on society.

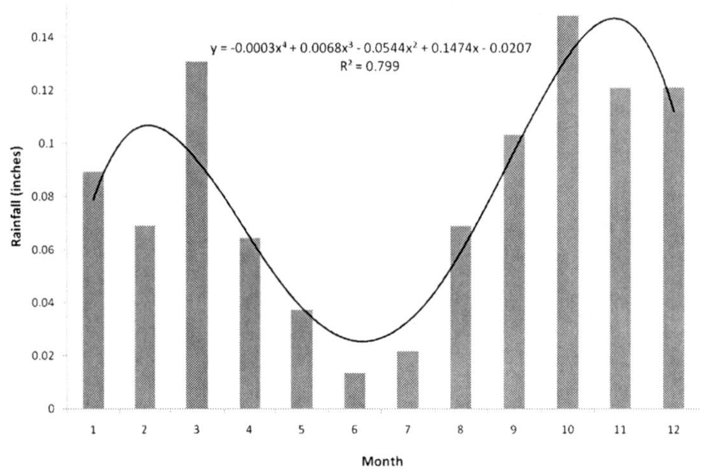

Figure 8. Fourth order polynomial fitting for the average rainfall data of Collin County, Texas, in 2006.

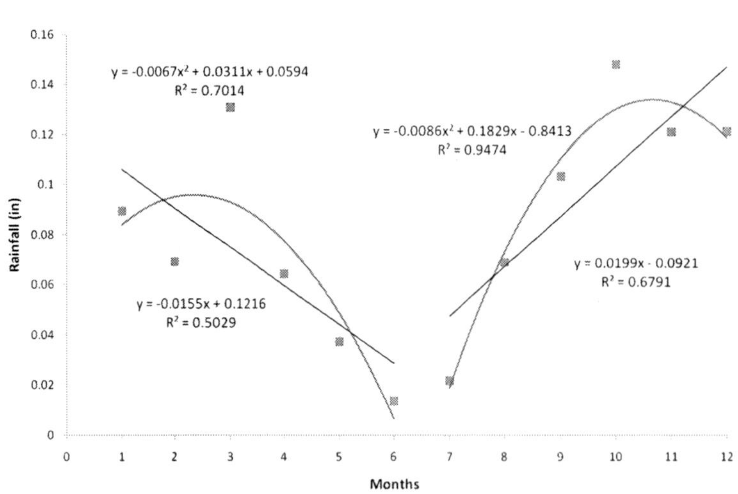

Figure 9. Piecewise linear and quadratic functions to fit average rainfall data for Collin County, Texas, in 2006.

EXTENSION

So far, the students' rainfall project has been limited to estimating rainfall using time averages. This type of model is too simple. After all, "average rainfall" does not mean that the same amount of rain falls everywhere in a selected area (state or county). A possible extension to this project-based learning (PBL) activity is to assume that measurements in each weather station (or gauge) represent the rainfall only in a given region around it. This PBL extension will improve students' understanding of spatial reasoning, geometry, and the concept of coordinate systems. The task of this PBL extension is to find the boundary lines that separate regions of influence and learn and understand the concept of weighted average (Lege, Bell, Davis & Nite, 2002).

Needed Terminology

Center of influence. Points representing an entire geographical area (the rainfall gauges).

Region of influence. The geographical area surrounding the center of influence. Each region of influence has a limit to how far it goes.

To illustrate the goals and procedures of this PBL extension, the data downloaded in the Explanation section will be used. That is, given the weather station positions in Collin County, Texas, the regions of influence and their boundaries will be estimated for this county. Table 2 shows the weather stations' latitudes, longitudes, and elevations, which were provided by NOAA (http://www. noaa.gov/).

Table 2. Geographic information about weather stations within Collin County

Gauge	***Station Name***	***County***	***Latitude***	***Longitude***	***Elevation***
1	Frisco	Collin	33:09:00	-96:48	740
2	Lavon Dam	Collin	33:02:00	-96:29	510
3	McKinney	Collin	33:14:00	-96:38	622

To perform this PBL extension, students need to follow the subsequent steps:

Step 1 □ Map Generation of a Region of Interest

Download and install Google Earth (http://earth.google.com/). In Google Earth, in the "Fly to" section, type Collin County, Texas, and press Enter. Google Earth will change the map view to zoom in on the Collin County geographical region. At this point, no boundaries will be shown. To enable boundary view, go to the "Layer" section and enable the "2nd Level Admin Regions (Counties)" checkbox. This will show colored lines on the boundaries of the counties. Now the zoom can be adjusted (with the wheel button in the middle of the mouse) to obtain a map of the whole county. The map should look like Figure 10 (without the yellow points).

Step 2 □ Location of Weather Stations

The coordinate system (latitude and longitude) from Table 2 is not the same as the one Google Earth uses. To convert this coordinate system to Google Earth's coordinate system and to locate the weather stations, go to the following website: http://www.earthpoint.us/Convert.aspx (without closing Google Earth). Now, type in the latitude and longitude values from Table 2 and click the button labeled "View Google Earth." You'll be asked to open or save a file for each coordinate searched. Select to open the downloaded file; doing so will pull up Google Earth, and a yellow point will be added to the current map, indicating the position of the weather station. Repeat this process for all weather stations in the selected region. A map similar to Figure 10 will be obtained.

Step 3 □ Map Scaling

Google Earth provides several tools to obtain map dimensions. One tool is the scale bar, located in the bottom left corner of the map. The other tool can be found by going to the Tools menu and selecting Ruler. This tool allows for drawing a line with the mouse and obtaining its dimensions. These tools are useful for estimating, for example, the area of the county.

Figure 10. Map of Collin County, Texas.

Step 4 □ Obtaining Regions of Influence and their Boundaries

Although this step can be done directly in the computer, it is recommended to print the map, use rulers and colored pencils, and follow the following steps:

- Draw line segments to connect the different weather stations (yellow lines forming a triangle in Figure 11).
- Draw boundary lines according to the following rules: A boundary line always passes through the midpoint of the line segment connecting two weather stations. A boundary line and the line segment between two weather stations are perpendicular (cyan lines in Figure 11).

Figure 11. Region of influence borders (cyan lines) and line segments connecting the weather stations (yellow lines).

By following these steps, students should obtain a map similar to Figure 11. In this case, three regions of influence were found. It is important to mention that more weather stations exist in Collin County; however, these stations do not provide information on the selected dates (from 2006 to 2008).

Step 5 □ Obtaining Weighted Averages

To obtain weighted average rainfall in Collin County, students should measure the percentage of area occupied by each region of influence. The area of a region of influence can be obtained using the Google Earth measuring tools explained in Step 3. Normally, the regions of influence have irregular shapes.

Therefore, it is useful to know that irregularly shaped areas can be calculated by employing subdividing, Pick's Formula, and Heron's Formula. The teacher should encourage students to do research about methods for calculating irregularly shaped areas. Once the areas of the regions of influence are calculated, they can be used to estimate the weighted average as follows:

Table 3. Weather stations' average rainfall and area of influence

Gauge	***Station Name***	***June 2007 Average Rainfall (in)***	***% Area of Influence****
1	Frisco	0.425	25%
2	Lavon Dam	0.429	30%
3	Mckinney	0.282	45%

*The areas of influence are only estimations.

For each weather station, multiply the measured rainfall value by the percentage of area occupied by region of influence. Then simply add these values to obtain the weighted average. Table 3 shows the average rainfall value for June 2007 and the percentage area of influence for each weather station.

By following the procedure listed above and using data from Table 3, the weighted average value obtained is 0.362 in. Comparing this value to the average of 0.379 in obtained by normal arithmetic proves that there is no significant difference between these two values. The teacher can start a discussion about the difference between weighted and arithmetic averages with the following question: What would be a possible reason for a greater difference between the values of these two types of averages?

This PBL extension employs geometry and technology to display a wide array of visual tools, which will help obtain more accurate conclusions. That is, students can not only state that a safe place to build a home is on the north region of Collin County (as concluded before; see the explanation section), but also set boundaries around this "safe region".

EVALUATION

Using different types of assessment such as formative and summative assessments is encouraged in this project. The use of a thinking/inquiry personal journal should be used in all project meetings; additionally, students will give a final oral presentation at the end of the project.

As mentioned by Barell (2007), journals are good for note taking during investigations. The process of recording and reflecting on the problem-solving situations students find themselves in offers opportunities for students to raise questions continually as they proceed through their inquiry. Moreover, their questions can be brought to the attention of the whole class. Table 4 shows some possible sentence stems for thinking/inquiry journals (Barell, 2007, p. 39). At any given time, the teacher can read students' curiosities and confusions. This will allow the teacher to identify whether their inquiries match their project predictions.

Table 4. Sentence stems for thinking/inquiry journals

What I wonder about now is…
What this means to me is…
This reminds me of (other ideas, concepts, experiences)…
What is important here is…
I feel…
What I'm learning is…
How this relates to my own experiences…

Assessment rubrics can be a particularly useful tool in assessing student presentations. The appendix contains shows the rubric for assessing students' oral presentations. This rubric is based on Iowa State University assessments (2005) and taken from Gallery Walk at Central Michigan University (http://serc.carleton.edu/files/introgeo/gallerywalk/oral_report1.v3.doc).

APPENDIX A

Oral Presentation Rubric

Criteria	Distinguished - 4	Proficient - 3	Basic - 2	Unacceptable - 1
Organization	Extremely well organized. Introduces the purpose of the presentation clearly and creatively. Student presents information in logical, interesting sequence, which the audience can follow. Ends with an accurate conclusion showing thoughtful, strong evaluation of the evidence presented.	Generally well organized. Introduces the purpose of the presentation clearly. Most information presented in logical sequence; a few minor points may be confusing. Ends with a summary of main points showing some evaluation of the evidence presented.	Somewhat organized. Introduces the purpose of the presentation. Student jumps around topics. Several points are confusing. Ends with a summary or conclusion; there is little evidence of content evaluation based on evidence.	Poor or nonexistent organization. Does not clearly introduce the purpose of the presentation. Presentation is choppy and disjointed; there is no apparent logical order to the presentation. Ends without a summary or conclusion.
Research Effort	Went above and beyond to research information, solicited material in addition to what was provided, brought in personal ideas and information to enhance project, and utilized more than eight types of resources to make project effective.	Did a very good job of researching; utilized materials provided to their full potential; solicited more than six types of research to enhance project; at times took the initiative to find information outside of school.	Used the material provided in an acceptable manner, but did not consult any additional resources.	Did not utilize resources effectively; did little or no fact-gathering on the topic.
Appropriate Use of Mathematical language: Grammar, Word Choice, Voice	Speaker uses clear articulation, proper volume, a steady rate, and shows enthusiasm and confidence. Poised and clearly comfortable in front of the group. Selects rich and varied words for context and uses correct grammar. Sentences are complete and grammatical, and they flow together easily. Mathematic and weather language is chosen for precise meaning.	Clear articulation but not as polished; slightly uncomfortable at times. Most can hear presentation. Selects words appropriate for context and uses correct grammar. For the most part, sentences are complete and grammatical, and they flow together easily. With a few exceptions, mathematic and weather words are chosen for precise meaning.	Audience occasionally has trouble hearing the presentation; speaker seems uncomfortable. Selects words inappropriate for context; uses incorrect grammar. Audience can follow presentation, but some grammatical errors and use of slang are evident. Some sentences are incomplete, halting, and/or vocabulary is somewhat limited or inappropriate.	Presenter is obviously anxious and cannot be heard or speaks in a monotone with little or no expression. Selects words inappropriate for context; uses incorrect grammar. Audience cannot focus on ideas presented because of the speaker's difficulty with grammar and appropriate vocabulary.

Use of Graphic Communication Aids	Graph is titled appropriately. All labels are correct and properly placed. All of the numerical data is correct, and the graph is proportional to the data. Graph appears proportionally without bias. Graph displays appropriate use of color or patterns and labels. Graph is very neat in appearance.	Graph is titled appropriately. Most labels are correct and properly placed. Most of the numerical data is correct, and the graph is proportional to the data. Graph appears proportionally without bias. Graph displays appropriate use of color or patterns and labels. Neatness could be improved.	Graph is not titled. Some labels are correct and properly placed. Some of the numerical data is correct, and the graph appears mostly proportional to the data. Graph appears mostly proportional with a small degree of bias. Graph displays color or patterns and labels; they are used in an appropriate way. Neatness needs much improvement.	Graph is not titled. Most labels are incorrect and improperly placed. Some of the numerical data is correct and the graph appears somewhat proportional to the data. Graph is not proportional and contains a glaring bias. Graph displays no color, patterns, or labels. Graph is not neat.
Eye Contact	Maintains eye contact, seldom returning to notes; presentation is like a planned conversation.	Maintains eye contact most of the time but frequently returns to notes.	Maintains some eye contact but at least half the time reads the report.	Student reads all or most of the report with no eye contact.
Audience Interaction	Involved the audience in the presentation; held the audience's attention throughout.	Presented facts with some interesting "twists"; held the audience's attention most of the time.	Some related facts, but went off-topic and lost the audience.	Incoherent; audience lost interest.
Length of Presentation	Within allotted time (5–8 minutes).			Too long or too short.

RELATED PROJECTS

Chambers, L. (n.d.). *Measuring local precipitation.* Retrieved June 16, 2009, from http://mynasadata.larc.nasa.gov/P7.php

Haddad, N. (2004). *Investigating the precipitation-streamflow relationship.* Retrieved June 17, 2009, from http://serc.carleton.edu/eet/module_discharge/ index.html

Spiegl, C. (n.d.). *Does cloud type affect rainfall?* Retrieved June 17, 2009, from http://mynasadata.larc.nasa.gov/L18_Spiegl.html

REFERENCES

Barell, J. (2007). *Problem-based learning: An inquiry approach* (2nd ed.). Thousand Oaks, CA: Corwin.

Lege, J., Bell, R. E., Davis, M., & Nite, S. (2002). *COMAP'S Mathematical models with applications.* New York: W.H. Freeman & Company.

Manduca, C., & Mogk D. (2002). *Using data in undergraduate science classrooms.* Retrieved June 21, 2009, from http://serc.carleton.edu/files/usingdata/UsingData.pdf

National Weather Services. (2009). *About NWS precipitation analysis.* Retrieved June 22, 2009, from http://water.weather.gov/about.php

Roberto Garcia
Mathematics Teacher,
Emmett J. Conrad High School, Dallas ISD

Alma F. Rangel-Chavez
Teaching, Learning and Culture,
Texas A&M University

JENNIFER STITT AND SERKAN ÖZEL

3. INTRODUCTION TO GEOGRAHICAL INFORMATION SYSTEMS

Recycle Project: What can you create out of all-recycled materials?

WELL-DEFINED OUTCOME

Students will create or engineer something completely from recyclable materials to become aware of "green" living and why they should recycle. At the conclusion of this project, students will develop a better connection to the earth and their living environment. They will develop an understanding of the role trash plays in everyday life. Students will gain knowledge on how to develop and build an object from simple products.

INTRODUCTION/BACKGROUND

Much of today's products are created from existing materials. Thus, today and in the future, engineers will create new products out of recycled items. At the start of the project, students go to a website to figure out their carbon footprint. As students explore the website, it shows them how many earths are needed to support their lifestyles. As an average, students use three earths over their life time to accommodate their lifestyles. Students' preparation for the project includes gathering items from home and school that are recycled. For instance, students can bring cardboard, paper, numbers 1–5 and 7 plastic, glass, metal, and so on. Students will need a large bag with their name to store the items in the classroom. Students will also need to rinse out any drink or food items. The teacher may want to suggest items such as cereal boxes, soda cans, food cans, and water bottles. Students should be given two weeks to collect items. The teacher should ask the students to collect more than they will need. At the end of the two week collection period, the teacher can either have the students create whatever they want or put some constraints on the end product. The teacher may decide to collaborate with teachers from different disciplines. For example, I am a Geographic Information Systems (GIS) teacher, and I collaborated with an English teacher who was reading a short story. Students that we shared created characters out of the story from all-recycled materials as an extension and then wrote about that character in English as part of their research junior research project. They designed musical instruments, computers, bridges, cars, and football fields. This project provided an opportunity for students to realize that used materials can be processed into new products similar to a brand new product.

The project takes several class periods and requires a lot of space to store items. The students will use X-Acto knives and/or tin cutters, so safety has to be discussed. This project can be implemented prior to teaching GIS software. The connection to the earth provides an interesting perspective to the students and requires them to learn software related to the earth and mapping. Additionally, the project prepares students for problem-solving and engineering activities they may face in the future. At the end of the project, students will be asked to write a report describing what they created, how it works, and how it will help society in the future. The students can be required to create projects that satisfy certain constraints. For example, the project has to (a) move people or things, (b) help save the environment, or (c) improve an existing technology.

OBJECTIVES

Mathematics

- Spatial reasoning plays a critical role in geometry; geometric figures provide powerful ways to represent mathematical situations and express generalizations about space and spatial relationships. Students use geometric thinking to understand mathematical concepts and relationships.
- Geometry consists of the study of geometric figures of zero, one, two, and three dimensions and the relationships among them. Students study properties and relationships having to do with the size, shape, location, direction, and orientation of these figures.

R. M. Capraro, M. M. Capraro, J. Morgan and J. Scheurich, (eds.), A Companion to Interdisciplinary STEM Project-Based Learning: For Teachers by Teachers, 27–34.

- Students analyze geometric relationships in order to make and verify conjectures.
- Students are expected to use constructions to explore attributes of geometric figures and make conjectures about geometric relationships.
- Students use tools to determine measurements of geometric figures and extend measurement concepts to find perimeter, area, and volume in problem situations. Students also find areas of regular polygons, circles, and composite figures.
- Students use numeric and geometric patterns to make generalizations about geometric properties, including properties of polygons, ratios in similar figures and solids, and angle relationships in polygons and circles.

Social Science

- Students understand the influence of scientific discoveries and technological innovations on daily life in the United States.
- Students use problem-solving and decision-making skills in a variety of settings, such as in independent and group work. Students are expected to use a problem-solving process to identify a problem; gather information; list and consider options, advantages, and disadvantages; choose and implement a solution; and evaluate the effectiveness of the solution. In addition, students are expected to use a decision-making process to identify a situation that requires a decision, gather information, identify options, predict consequences, and take action to implement a decision.
- Students are expected to identify a source, its use, its quality, and conservation of water as well as document the use and conservation of both renewable and non-renewable resources.

Language Arts

- Students use recursive writing processes when appropriate. Students are expected to use prewriting strategies to generate ideas, develop voice, and plan.
- Students are also expected, both individually and collaboratively, to develop drafts by organizing and reorganizing content and by refining style to suit occasion, audience, and purpose.
- Students are expected to proofread writing for appropriateness of organization, content, style, and conventions.
- Students need to refine selected pieces in order to tailor their writing for general and specific audiences.
- Students should use technology for aspects of creating, revising, editing, and publishing texts.

Science

- Students are expected to make responsible choices in selecting everyday products and services, using scientific information, and evaluating the impact of research on scientific thought, society, and the environment.
- Students are expected to make sound judgments about the use and conservation of resources and disposal or recycling of materials.
- Students will evaluate the impact of waste management methods such as reduction, reuse, recycling, and composting on resource availability.
- Students will research the advantages and disadvantages of "going green" such as organic gardening and farming, natural methods of pest control, hydroponics, xeriscaping, energy-efficient homes and appliances, and hybrid cars.

GIS

- Students are expected to know and understand contemporary issues in the context of spatial and environmental perspectives.
- Students are expected to identify renewable and non-renewable resources that must come from outside an ecosystem, such as food, water, lumber, and energy.
- Students are expected to analyze and evaluate the economic significance and interdependence of components of the environmental system.

MATERIALS NEEDED

- Scissors
- Tin cutters
- Hot glue guns
- Glue sticks (approximately two for each student)
- Regular glue
- Duct tape
- String
- Computers
- Web access
- Microsoft Word
- Microsoft PowerPoint
- Rulers
- Printer
- Math Chart (volume, circumference, etc.)
- Recycled materials (plastics 1–5 and 7, cardboard, paper, glass, food containers)
- Paint (students may be allowed to paint their creations; however, paint tends to hide the materials that were used.)

STUDENT INTRODUCTION

Posters or flyers about recycling and items students can recycle can be put up around the school. The teacher may encourage students to look for the posters in order to generate interest for the project. Students will distribute a survey around school. The survey should contain the following questions:

- Do you recycle?
- Do you know where to look for the recycle number on plastic?
- Do you know what "biodegradable" means?
- Do you think it is important to recycle?

Students need three days to collect data. Then, they will graph the data on either the board or a computer using bar graphs. These graphs provide an overall picture of the school demographic as well as providing a discussion opportunity about recycling awareness in the school. The next step for students in the project is the carbon footprint exercise. Following the carbon footprint exercise, the teacher distributes the handout entitled "Engineer an Object out of Recycled Materials." In this activity, students have to design an object using recycled materials. Objects they design will have three constraints. Their project must meet these constraints:

- Help the planet
- Move people or things
- Improve on an existing technology
- Use only recyclable materials
- Use only the supplies provided in class

After students finish their project, they will present how they engineered an object out of recycled materials. Students should be collecting recycled items several weeks before the start of the project.

ENGAGEMENT

The teacher begins the project with an open discussion about sustainability, recycling, biodegradable items, and anything "green." The teacher discusses the importance of using biodegradable items for everyday needs. Following the discussion, the teacher directs students to conduct research for approximately 15 minutes about biodegradable items and report their findings back to their group. Student findings may include the meaning of "biodegradable" and sample materials that are and are not biodegradable. Then the teacher asks students if they recycle at home and facilitates a discussion about whether the school should recycle. After the discussion is over, the teacher introduces the carbon footprint website (http://www. footprintnetwork.org). This website contains a quiz with questions that assess students' daily habits (see Figure 1). Each student must complete the quiz individually. As a result of the quiz, students will receive information about the number of earths needed to sustain their lifestyle. Because there is only one earth, any quiz result showing the need for more

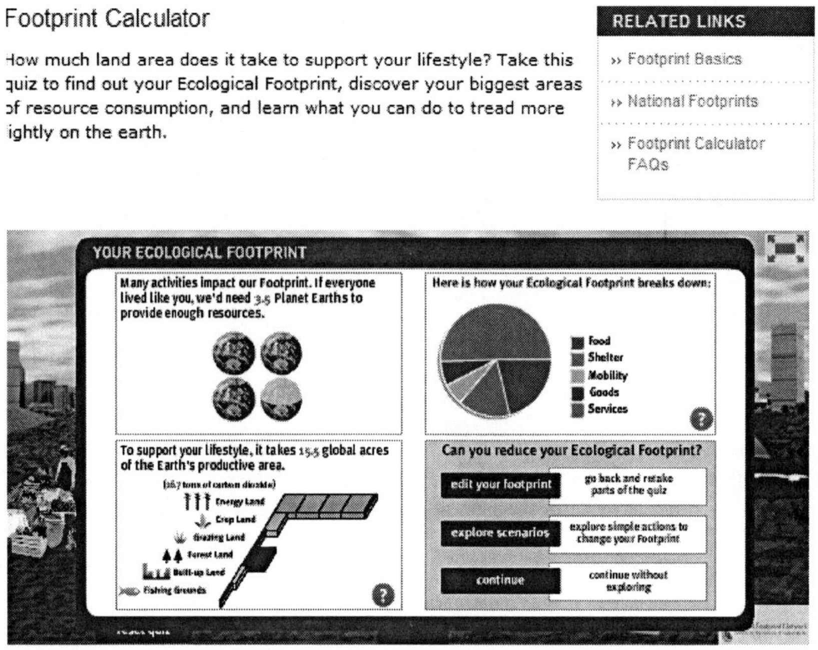

Figure 1. A screen shot of typical results of the carbon footprint website quiz.

than one earth will be striking and lead to a discussion about what should be done to lower their carbon footprint: recycle! Then the teacher can segue into the project. The carbon footprint exercise takes about 40 minutes.

EXPLORATION

Students will examine their materials and read the constraints in order to have a better idea of what they are going to create. This process requires brainstorming, which should be documented. Students may spend a couple of class periods brainstorming about their design. The teacher should emphasize the importance of measurement and patience in this project and points out the effectiveness of their brainstorming. Then students can do preliminary drawings of what one of their objects might look like. If they want items to look realistic, they may need to find a picture and decide how they will create a scale model of the project. The teacher, in this step, must talk about safety and the system for distributing supplies. The teacher may need to use a checkout system for dangerous supplies like X-Actor knives. The students will develop a design notebook where they will keep their designs for their object. This design will be the basis for part of their grade. They are allowed to make changes to their design, but all changes must also be reflected in their design notebook along with specifications for dimensions and assembly order and the materials for assembling each piece.

EXPLANATION

Students have had an open classroom discussion about biodegradable items, recycling, and the terminology “carbon footprint” After the discussions, the students will log onto computers and go to the website http://www.Footprintnetwork.org. There are other websites available, and they can be located easily by searching for “carbon footprint quiz.” For this particular website, students should click on the link “Footprint for you.” This link will lead students to a quiz that asks them several different questions about what size house they live in and how they get around from place to place.

Students’ scores on the quiz can be printed out and hung on the classroom wall so students can compare their own carbon footprints with their peers. The goal of this quiz is to have less and not more. The quiz explains why recycling and reusing is important. After the carbon footprint exercise, the teacher can pass out the student handout.

Next, the teacher informs students that they will be creating an object out of the items they collected given the constraints. The teacher can either use the constraints provided in this chapter or create a new set of constraints. For example, this project can include geometry objectives related to shapes, surface area, altitude, and volume.

EXTENSION

As an extension to this project, if students are discussing or reading about a certain character in English or social studies, they can use recycled materials to design an author or character they are learning about. In this case, students may paint the objects to better reflect the description of the character in the book or author. As an additional extension, in mathematics, students can be given miniature representations of objects then required to make larger scaled versions of the items. The teacher can vary the difficulty of the task by assigning various scale factors that increase or decrease difficulty. For inclusion students, they can be assigned 1:1 scale but required to develop a pedestal on which their object will rest without having an object to copy.

To modify the present activity the teacher can have students work in groups to research closely related topics such as receptacle containers for recyclable materials and one group can be given the task of juice containers (rectangular prisms), coke cans, and for added difficulty cardboard boxes and paper. The challenge with the last two is the balled up paper takes up more room than in sheets as does cardboard boxes (if they remain in box form and are not broken down). By making the tasks closely related it enhances the synergistic cooperation because they can learn from each other without having to compete to create the best copy.

EVALUATION

Appendix A is an example rubric for the student presentations. Appendix B is a sample rubric for the recycled item project.

APPENDIX A

Presentation Rubric - 16 points

	Exemplary 4	*Mastery 3*	*Proficient 2*	*Introductory 1*
Visuals	Highly developed, memorable visuals. Illustrate main points. Visuals are integral to the main points and extremely well organized.	Visuals clarify and illustrate main points. Visuals are useful, applicable, neat and organized.	Visuals clarify and illustrate main points. They are tangentially related to the main points. Mostly neat, organized, and applicable.	Visuals are not tightly linked to presentation and do not support or clarify main points.
Communication of Main Points	Communication is articulate, easily heard, and well thought out. Main points are highlighted through intonation, gesture, or visuals.	Communication is clear; majority of the main points are covered and mostly in logical sequence.	Communication is somewhat lacking, but main points are still evident.	Communication is poor, and main points are not fully covered.
Posture, diction, & confidence	Student stands tall, shows confidence, speaks authoritatively, and can answer questions about the project.	Student mostly stands tall, shows confidence, speaks authoritatively, minor difficulty answering questions	Student is nervous or unsure, relies on notes or reads directly from slides but conveys main points of the project.	Student is unsure of what to say or what comes next and has difficulty answering questions.
Fluency	Speech is articulate and flows perfectly. Student does not use filler words.	Student does not stumble over words, but uses a few filler words.	Student has difficulty speaking fluently or uses filler words but perseveres to communicate effectively.	Student consistently stumbles over words and uses a lot of filler words.

APPENDIX B

Recycled Item Rubric - 16 points (counts twice = 32 points)

	Exemplary 4	*Mastery 3*	*Making Progress 2*	*Introductory 1*
Required Elements	All materials were recycled or biodegradable and contained on the materials list.	More than 75 percent of the materials were recycled or biodegradable.	More than 50 percent of the materials were recycled or biodegradable.	Less than 50 percent of the items were recycled or biodegradable.
Meeting Constraints	The object satisfied all the constraints.	The object satisfied most constraints.	The object satisfied some important constraints.	The object was incomplete or did not satisfy most constraints.
Attractiveness	The object was visually interesting, clean and neat. Labeling was accurate and complete.	The object was visually interesting but with minor issues of cleanness and neatness. Labeling was incomplete or inaccurate.	The object contained visually appealing elements but some sloppiness was evident. Labeling was poor, sloppy, and or incomplete.	The object was visually unappealing, sloppy, or appeared to be falling apart. Labeling was mostly inappropriate or missing.
Project Engagement	Student was actively engaged during the project, enthusiastic and cooperative with peers and teacher.	Student was mostly engaged in the project, mostly enthusiastic and cooperative with peers and teacher.	Student was partially engaged in the project, had to be redirected occasionally.	Student was off-task for most of the project.

EXAMPLES OF STUDENT WORK

Figure 2. A sample of student work.

Figure 3. A sample of student work.

RESOURCES/INTERNET SITES

www.footprintnetwork.com
http://www.foodproductiondaily.com
http://www.friendlybags.com/facts_and_effects.php
http://www.genpak.com/cfm/biodegradable.cfm/

Jennifer Stitt
GIS Teacher,
Emmett J. Conrad High School, Dallas ISD

Serkan Özel
Educational Psychology,
Texas A&M University

KATHY VONDRACEK

4. THEORY OF BIOLOGICAL EVOLUTION

A creative approach for teaching an important and sometimes controversial topic

WELL DEFINED OUTCOME

Students will work in teams to explore the process of evolution and to apply research to a new species.

INTRODUCTION

The theory of evolution is an important content area in biology. Students will have been exposed to several biological concepts prior to the time this project is assigned. Topics they will further explore include: classification of organisms, evolution of prokaryotes and eukaryotes, fossil records, extinction, genetic variation, and the naming of species. Charles Darwin, a naturalist, sailed on a voyage in 1831 from England to South America and then to the Galapagos Islands and back to Europe. He collected data on a variety of specimens and fossils of animals, and he noticed that there was similarities in both plants and animals between those found on the Islands and the ones found in certain places in South America. Darwin is recognized for his theory of evolution on the basis of natural selection. The definition of natural selection is summarized, as the process by which species that possess favorable variations and are better adapted to their environment will have a better chance to survive and produce offspring than those that have less favorable adaptations to their environment.

OBJECTIVES

Biology and Scientific Process

- Using scientific evidence and information, the student will analyze, review, and critique scientific explanations, including hypotheses and theories, as to their strengths and weaknesses.

Biological Evolution

- The student will identify evidence of change in species using fossils, DNA sequences, anatomical similarities, physiological similarities, and embryology.
- The student will also illustrate the results of natural selection in speciation, diversity, phylogeny, adaptation, behavior, and extinction.

Applications of Taxonomy

- The student will analyze relationships among organisms and develop a model hierarchical classification system based on similarities and differences using taxonomic nomenclature.

Interactions and Interdependence that Occur Within an Ecosystem

- The student will compare variations, tolerances, and adaptations of plants and animals in different biomes as well as identify and illustrate that the long-term survival of species is dependent on a resource base that may be limited.
- The student will interpret interactions among organisms exhibiting predation, parasitism, commensalism, and mutualism and understand food chains and food webs.

R. M. Capraro, M. M. Capraro, J. Morgan and J. Scheurich, (eds.), A Companion to Interdisciplinary STEM Project-Based Learning: For Teachers by Teachers, 35–41.

Social Science

- The student will analyze geographic data to determine what species live in specified areas of the world.

Language Arts

- The student will be required to organize ideas in writing to ensure coherence, logical progression, and support for ideas.
- The student will use prewriting strategies to generate ideas, develop voice, and plan.
- The student will compile written ideas and representations into reports, summaries, or other formats and draws conclusions.

MATERIALS

- Computers
- Resources such as textbooks, history books, and the Internet
- Magazines to cut pictures out of
- Poster boards
- Paper
- Markers
- Map pencils
- Rulers
- Construction paper
- Nnotebook paper
- Computer paper
- Microsoft PowerPoint jump drive to save work

STUDENT INTRODUCTION

Dr. Oliver Schwimmer is a renowned marine biologist. He has travelled to many places throughout the world collecting data on different aquatic flora and fauna. He has discovered what he believes is a new species of plant that seems to be adapted to land as well as an aquatic environment. What mystifies him is how the plant got away from the water's edge. It is a type of seaweed that is found below the surface of relatively shallow waters. The plant's habitat is salt water, and it can usually be found near mangrove swamps. Dr. Schwimmer's first discovery was in Jamaica, and now he thinks he may have found a similar plant on the island of Aruba. He wants to form a research team of scientists to help him explore the world and try to find out if this plant is found anywhere else. The team also needs to explore what possibly feeds on the plant and how it is able to exist outside the aquatic environment it derived from. Typically, many aquatic organisms feed off of these plants and use them for shelter to hide from predators. Dr. Schwimmer needs you to join his team and help choose the other professional members. Through research and exploration with Dr. Schwimmer, the team discovers a new species of its own. What type of professional doctors should be included? What will the mode of travel be for the exploration? Will there be any published documents about the research done by the team? What is the time frame for the team to collect the research? During what time era will the team's research take place? The dates play an important role in the outcome of the project because you must consider the technology that was present during the era selected.

There will be an oral presentation done by the research team at the end of the project. You will be given rubrics to assess your day-to-day progress and the final presentation.

Good luck!

ENGAGEMENT

The teacher starts off the project by having an open discussion about evolution. The teacher should create dialogue to develop the students' interest in the possibility of discovering a new species. Students will choose the specie they would like to investigate in depth. Students will seek answers to questions such as; - What will the animal and/or plant that they decide on look like? - What ancestral links does their new species have?

The teacher allows for some group brainstorming to take place; students should use the time to write down all of their ideas. The teacher asks the students to write down anything they think they will need in order to get started on the project. Once the students turn in their papers, the teacher lists some of the items the students wrote down on an overhead projector.

The teacher starts another discussion about the types of researchers that will be invited to join the research team. Students will each choose a particular profession and report back to the group what their specialty of science is and how this will benefit the team. They will also investigate where they will travel in order to do their research. The engagement and initial activities will take two class periods.

EXPLORATION

The next time the teacher meets with the students, he or she should present the story in the Student Introduction above as well as the story of Darwin's exploration. Next, the teacher will ask the students to decide where they plan to travel using the Internet, textbooks, or library resources and what role each one of them will play during the exploration.

Each student will be responsible for his or her affiliation to the group and for informing the team what role he or she plans on playing in the research. For example, what do botanists study and how would they be beneficial to the group? Students will have to write down items they need in order to start their exploration. They will also need to start thinking in terms of their own theory of evolution. Do they want to compile a new theory or choose to adopt Darwin's theory? They can adapt his theory to fit their own beliefs.

EXPLANATION

Now the students should understand that their task as a research team is to help Dr. Schwimmer. After they have done some initial research on the Internet about evolution, Darwin, biological terminology, and different professionals, students can begin to develop their own concepts and research as to what their new species will be and how to present it. Each student needs to work individually at first class. Later on, they need to spend at least 20–30 minutes of each class meeting to plan as a team, make sure no one is off task, and decide what materials each team member will contribute.

The students on each team will need to rotate the responsibility of writing out a brief status report to turn in to the teacher at the end of each meeting period. The teacher needs to explain that this status report will help students keep their focus on the purpose of their research and manage their time effectively during class. The report allows teacher insight as to where the students might need assistance during the next class meeting.

Students should be aware of what materials they will need to complete the project. If Google Earth or Google Maps is not available, they can create their own maps using materials such as poster board, map pencils, markers, etc. Rubric 1 (see Appendix A) will be used for student assessment in the map section of their projects.

Examples and ideas about their research to be given to the students are listed below. Teacher may choose to talk about these examples anytime during their investigation phase.

- In what type of habitat/environment did you find your species?
- What does it eat and what eats it? (Predator/prey)
- What did your species evolve from? Show pictures/fossil records, etc.
- What type of plants/flora was found in the environment where you travelled and located your species?
- Does your species have a special instinct or ability to survive?
- Fossil records that indicate what the specie evolved from,
- Is it closely related to any living thing?
- Is it a carnivore, omnivore, herbivore, or autotroph?
- What is the closest type of species to yours and where is it found?
- Include any other relative information; you should be creative with this project!

Students should also give background information about their group, like:

- How did you meet or team up?
- How long was your journey together (include dates)?
- Include what type of transportation was used: boat, plane, driving, walking/hiking, etc.
- Did you publish your journal or a book about your travels?
- Include a map, timeline, and pictures of your journey together,

- List the places you travelled to,
- Make sure you use biological terminology throughout your entire presentation.

The presentation must have pictures of the students' species and the evolutionary process of the species. The pictured presentation should include a map, but also may include a time line, or a combination of everything.

The teacher should share and explain Rubric in Appendix B with the students.

EVALUATION

The rubrics are used to help students guide their learning. Rubrics are used as both summative and formative evaluations. Students have the experience of being rated both by someone using the rubrics as well as using the rubrics themselves. When they spend the time to rate each other, it improves their work. They should reflect on each rubric before it is used to provide formal feedback and grading both by someone using the rubric as well as by using the rubrics with each other (See Appendix A, B, and C for sample rubrics).

EXTENSION

Microsoft PowerPoint is an added feature and will be considered an extension, but it should not take the place of an actual poster or artifact of the project species. If there are interested students who would like to learn more about evolution, or particularly about Darwin, they can be assigned to research his life and present their findings to the rest of the class. They can investigate the influence of Darwin's theory on other social and scientific areas. Some of Darwin's books are free to download at the following links:

- *The Origin of Species*
  http://books.google.com/books?id=TCwLAAAAIAAJ&printsec=frontcover&dq=Charles+Darwin
- *The Voyage of the Beagle*
  http://books.google.com/books?id=HalyPrin8MAC&printsec=frontcover&dq=Charles+Darwin

APPENDIX A

Map rubric

	Exemplary 4	Mastery 3	Proficient 2	Introductory 1
Required Elements	Map contains a North Arrow, Legend, Scale, and Title. Title is catchy and layout looks exemplary. Map is zoomed to the correct extent.	North Arrow, Legend, Scale, and Title are present. Map is zoomed to correct extent.	Something is missing, either the North Arrow, Legend, Scale or Title. Map is not zoomed to the correct location and may not be zoomed in to the correct extent.	Several required elements were missing
Labels	Elements included on map are labeled. This includes roads, cities, states, etc. Only the elements used are the elements labeled. Labels are original and are not the default label.	Elements included on map are labeled. This includes roads, cities, states, etc. Only the elements used are the elements labeled.	Many elements on the maps that are included are not labeled and labels are NOT original.	Labels do not exist or do not match the elements on the map.
Graphics - Relevance	Extra graphics were used, such as inset maps, charts or graphs etc and are placed well on the layout.	Extra graphics were used, such as an inset map, charts or graphs, but they might not have been placed in the most relevant of places.	An attempt to use additional graphics was made, but did not end up on the final layout.	No additional graphics were used.
Attractiveness	The map is neat and free of any spelling errors. Legend is visible and understandable. Map colors go well with the defense of the map. Student went above and beyond to make the map look professional and did not add any cartoons or weird graphics to try and make it look better.	The map is neat and free of grammatical errors. Colors were used wisely.	The map is somewhat neat, but seems to be confusing to understand.	The map does not have any required elements and is not neat and professional.
Grammar	There are no grammatical/mechanical on map.	There are 1-2 grammatical/mechanical mistakes on the map	There are 3-4 grammatical/mechanical mistakes on the map.	There are more than 4 grammatical/mechanical mistakes on the map.

APPENDIX B

Participation Rubric

	Exemplary 4	Mastery 3	Proficient 2	Introductory 1
Attendance	Student is present all or most days.	Student missed a few days, but made up work.	Student missed several days and did not make up all days.	Student was absent most of project.
Work Ethic	Student is always on task and willing to ask questions when needed.	Student is on task most of the time.	Teacher has to ask student more then once to stay on task.	Student is always off task and has to be asked numerous times to be quiet and complete their work.
Effort	Student puts in effort most if not all class period. Student does not pack up and wait at door for bell to ring.	Student puts in a lot of effort, but has been off task once.	Student puts in little effort and is off task several times in the class. Student packs up early.	Student is always off task and packs up early every day.
Attitude	Student is positive!	Student is positive most of the time.	Student is not very positive but is still trying.	Student is not positive and disrupts class.
Group Work	Student is productive when working with a group. Student actively participates.	Student is productive, and only slacks off once.	Student sometimes makes the other people in the group do all the work.	Student only delegates and does not work well with the group.

APPENDIX C

Summative Rubric

CATEGORY	*4*	*3*	*2*	*1*
Collaboration with Peers	Almost always listens to, shares with, and supports the efforts of others in the group. Tries to keep people working well together.	Usually listens to, shares with, and supports the efforts of others in the group. Does not cause "waves" in the group.	Often listens to, shares with, and supports the efforts of others in the group but sometimes is not a good team member.	Rarely listens to, shares with, and supports the efforts of others in the group. Often is not a good team member.
Content	Shows a full understanding of the topic.	Shows a good understanding of the topic.	Shows a good understanding of parts of the topic.	Does not seem to understand the topic very well.
Vocabulary	Uses vocabulary appropriate for the audience. Extends audience vocabulary by defining words that might be new to most of the audience.	Uses vocabulary appropriate for the audience. Includes 1–2 words that might be new to most of the audience, but does not define them.	Uses vocabulary appropriate for the audience. Does not include any vocabulary that might be new to the audience.	Uses several (5 or more) words or phrases that are not understood by the audience.
Enthusiasm	Facial expressions and body language generate a strong interest and enthusiasm about the topic in others.	Facial expressions and body language sometimes generate a strong interest and enthusiasm about the topic in others.	Facial expressions and body language are used to try to generate enthusiasm, but seem somewhat faked.	Very little use of facial expressions or body language. Did not generate much interest in the topic being presented.
Speaks Clearly	Speaks clearly and distinctly all the time (95–100%) and mispronounces no words.	Speaks clearly and distinctly all the time (95–100%) but mispronounces one word.	Speaks clearly and distinctly most of the time (85–94%). Mispronounces no more than one word.	Often mumbles or cannot be understood OR mispronounces more than one word.
Comprehension	Student is able to accurately answer almost all questions posed by classmates about the topic.	Student is able to accurately answer most questions posed by classmates about the topic.	Student is able to accurately answer a few questions posed by classmates about the topic.	Student is unable to accurately answer questions posed by classmates about the topic.
Research	Worked diligently and efficiently every day to prepare for the overall presentation and supported others on the team.	Worked well most days and made good use of time. Usually helped others with the project.	Had to be redirected to get back on task a few times. Was not as helpful as others on the team and required their assistance to complete the project.	Did not work efficiently and was usually off task. Did not prepare a portion of the research assignment and did not demonstrate a team effort.

SUPPORTING AREA CONTENT

Examples of Student Work

Figure 1. Student project.

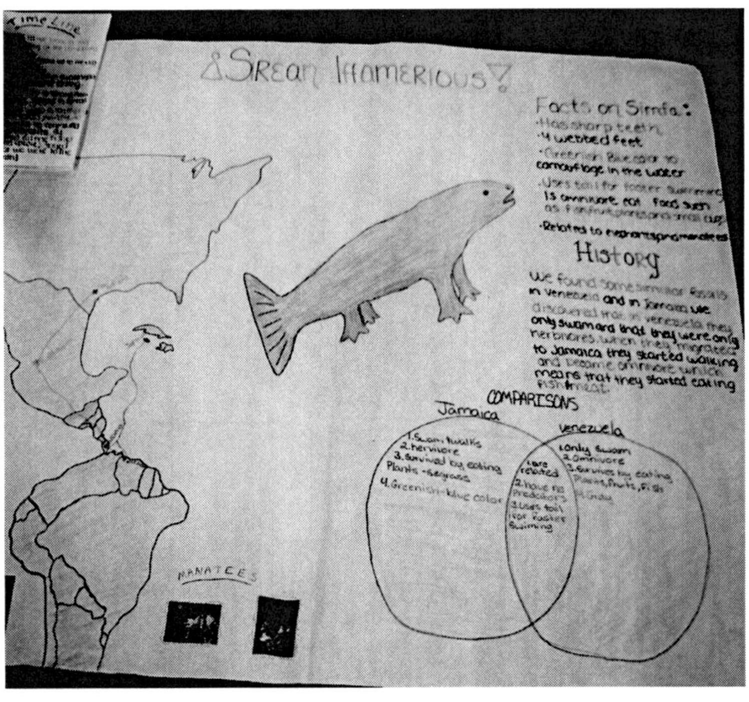

Figure 2. Student work.

Kathy Vondracek
Chemistry Teacher,
Emmett J. Conrad High School, Dallas ISD

JENNIFER STITT AND SERKAN ÖZEL

5. INTRODUCTION TO AUTOCAD: SAY GOODBYE TO WINTER

How would you design a snowflake in a two-dimensional environment?

WELL-DEFINED OUTCOME

At the conclusion of this project, students will have an understanding of several important AutoCAD and geometric concepts. These concepts include, but are not limited to, symmetry, mirroring, scale factor, distance, midpoints, and endpoints.

INTRODUCTION/BACKGROUND

This is a one to three day project introducing fundamental concepts in AutoCAD while addressing symmetry. The duration of this project will depend on the number of snowflakes students design. Students should already have basic knowledge of AutoCAD before they start the project. If students lack AutoCAD knowledge or if they do not have access to AutoCAD, the teacher can have students either use some other two-dimensional design software or draw on graph paper. The students will draw different versions of snowflakes using a very simple step by step concept that originates from a hexagon.

OBJECTIVES

Mathematics

- The student uses geometric thinking to understand mathematical concepts and relationships. Spatial reasoning plays a critical role in geometry. Geometric figures provide powerful ways to represent mathematical situations and to express generalizations about space and spatial relationships.
- The student uses geometric thinking to understand mathematical concepts and relationships.
- The student finds areas of sectors and arc lengths of circles using proportional reasoning.
- The student uses one- and two-dimensional coordinate systems to represent points, lines, line segments, and figures.

AutoCAD/Drafting/Technology Applications

- The student is expected to gain problem solving skills and group work skills by seeking and responding to his or her peers and professionals in delineating technological tasks and problem solving.
- The student is expected to know the concepts and skills that form the technical knowledge of computer-aided design.
- The student is expected to customize screen menus to fit specific problems or needs.
- The student is expected to construct points, lines, and other geometric forms using accepted, computer-aided design methods and create a freehand, simple one-point perspective.

MATERIALS NEEDED

- Computer
- AutoCAD or other two-dimensional design software
- Graph paper (if students do not have access to AutoCAD or any other two-dimensional design software)
- Printer or plotter (color)

R. M. Capraro, M. M. Capraro, J. Morgan and J. Scheurich, (eds.), A Companion to Interdisciplinary STEM Project-Based Learning: For Teachers by Teachers, 43–48.

STUDENT INTRODUCTION

No two snowflakes in the world are alike. It can be fun to create a snowflake. However, there are two basic rules to follow: (a) Snowflakes are symmetrical, and (b) snowflakes have six arms. If students are going to use AutoCAD, the following settings should be adjusted. Turn on "osnaps" and "midpoint snap" and set the "polar snap" to 15 degrees. Also, turn off the "grid" and "snap." If students are using different two-dimensional design software, they can find similar features and adjust them.

ENGAGEMENT

The teacher starts class with a discussion about snowflakes. The teacher guides students to talk about the shapes of snowflakes and asks if they think all snowflakes are the same. Other important discussion points are the arms and the symmetrical feature of snowflakes. Students can be engaged into discussion by asking about their experiences with snowflakes.

EXPLORATION

After the engagement period, ask students to work on AutoCAD to draw symmetrical figures. Students can try to explore AutoCAD to draw symmetrical figures and a snowflake without any instruction. After students have finished exploring AutoCAD, the teacher hands out the student introduction with instructions.

EXPLANATION

Students need to follow their instructions and adjust the settings in AutoCAD according to the student introduction. To draw snowflakes:

- Students need to draw a hexagon, which is a six-sided polygon, by selecting "polygon" in AutoCAD. The size of the hexagon can be left to student choice (see Figure 1).

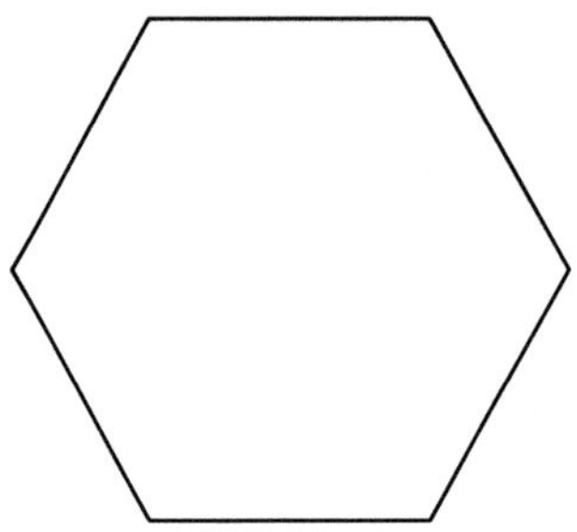

Figure 1. A hexagon.

- Students need to select the "line" command and draw lines from each corner to the opposite corner (see Figure 2). Enabling "snaps" makes this process easy.

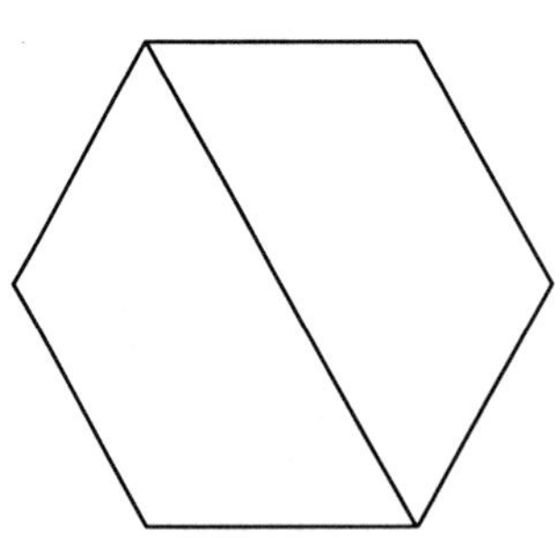

Figure 2. A hexagon with a line from one corner to the opposite corner.

- The student can repeat the second step to create a figure similar to Figure 3. However, AutoCAD has a symmetry features to accomplish this objective. The student can use the "mirror" button to create other lines from one corner to the opposite corner. The student clicks the "mirror" button, selects line, and hits "Enter" on the keyboard. When AutoCAD opens a pop-up menu asking where the "mirror line" should begin, the student clicks the middle of the existing diagonal line. When the student rolls his or her mouse over the middle of the diagonal line, the cursor turns into a triangle. The student should select this middle point and then look for the next point of the "mirror line." If the "polar snap" is set to 15 degree intervals, the line will rotate around the middle point, and eventually the student will see the line appear in opposite corners. Then, the student releases the mouse button to set the line as it appears on the screen. The following question will pop up when the student releases the mouse button: "Erase the source object?" The source object should be kept in order to have all lines appear on the hexagon. The student should repeat this step until the hexagon looks like the figure in Figure 3.

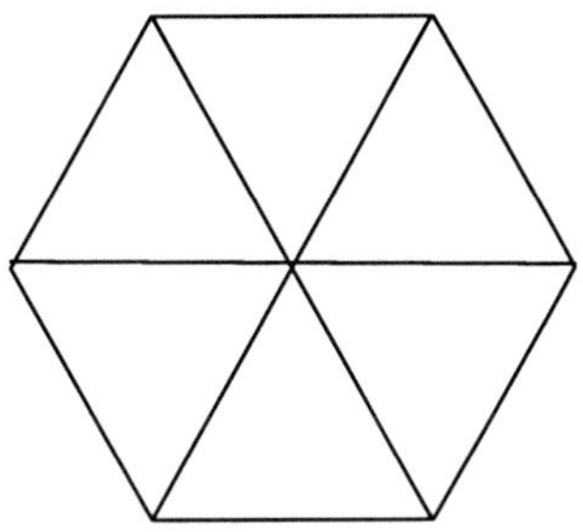

Figure 3. A hexagon with lines from corner to corner.

- The hexagon is a benchmark for the arms of a snowflake. In this step, students will remove the hexagon and leave the lines drawn in the first and second steps. The new shape will look like Figure 4.

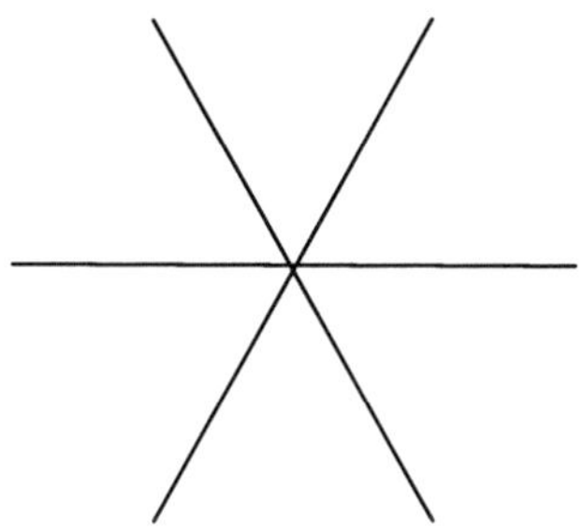

Figure 4. A snowflake base.

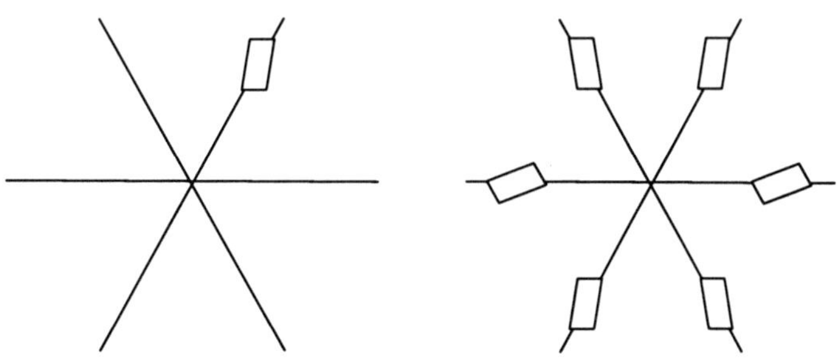

Figure 5. Creating a snowflake by adding figures using symmetry.

- After the base of the snowflake is created (see Figure 4), students will build their own snowflake on this base. Snowflakes are symmetrical, thus, any shape drawn on one side of an arm should be drawn on the other arms, to make them symmetrical to the shape on the other side of the line of symmetry (see Figure 5).

Students will use the "line" command and "mirror" button to accomplish this step. The teacher can encourage students to start with a simple shape and see the finished snowflake before they proceed to more complex figures. The snowflake in Figure 5 is a basic snowflake.

- Students are limited only by their imagination. Students are encouraged to use their imagination to create their own original snowflakes. The teacher can ask students to design at least five snowflakes. Each snowflake should be more complex than the previous one. If students do not have two-dimensional design software like AutoCAD, they should draw snowflakes on graph paper. To create symmetrical snowflake arms on graph paper, students may fold their paper as well as use geometry. Snowflake will be evaluated using the rubric in the appendix.

EVALUATION

The rubric included in Appendix A can be used to evaluate the created snowflake by assessing the required elements, attractiveness, and AutoCAD elements.

EXTENSION

As an extension to this project, students will have a choice of digitizing either a cartoon face or a rim for a car. Because students had to focus on symmetry and mirroring in order to draw the snowflake, a rim or cartoon face will emphasize the same concept. Students will do their own research via the Internet to get ideas about cartoon faces and rims. Students will also be asked to find symmetry and mirroring without using software.

APPENDIX A

Snowflake Rubric □ 12 points

	Exemplary 4	*Mastery 3*	*Proficient 2*	*Introductory 1*
Required Elements	Snowflakes have 6 arms and are created from a hexagon. Snowflakes increase in complexity as the student progresses.	Snowflakes have 6 arms. Symmetry is used well. They increase in complexity somewhat.	Snowflakes are present but not all have 6 arms. Snowflakes do not increase in complexity.	Not all snowflakes are present. Not all have 6 arms. There is a lack of attention to detail.
Attractiveness	Snowflakes are attractive and look intricate. Attention to detail is present.	Snowflakes are attractive. Some look intricate and some attention to detail is present.	Snowflakes are sloppy but present. There appears to be a lack of attention to detail.	Snowflakes are poorly designed and are not intricate.
AutoCAD Elements Used	Student used all AutoCAD commands, such as Mirror, Snap, and Polar Angles.	Student used several AutoCAD commands.	Student used some AutoCAD commands but also drew freehand.	Student did not use AutoCAD commands and did everything freehand.

EXAMPLES OF STUDENT WORK

Figure 6. An example of student work.

Figure 7. An example of student work.

Jennifer Stitt
GIS Teacher,
Emmett J. Conrad High School, Dallas ISD

Serkan Özel
Educational Psychology,
Texas A&M University

LEE WHITEHEAD

6. SIX KINGDOMS OF LIFE

What does the world look like to you?

WELL DEFINED OUTCOME

Students will relate the classification of living organisms to a real world classification task, such as organizing music, and display this information in a book format categorizing the six kingdoms of life.

INTRODUCTION/BACKGROUND

All organisms on the planet can be placed into one of the six kingdoms of life that scientists have created to help organize the world around us. This project will help students learn to research and organize information they have gathered, just like they would classify and categorize their music. This project is generally done near the end of the year to help reinforce the skills the student have previously learned during the year. The idea behind this project is to help create a student-led project covering the classification of life into the six kingdoms. The information students gather will be placed in a book format with one page for each of the six kingdoms. The last page will be used as a word bank; this will be a place for students to include science vocabulary, which will be listed and defined to help others who read the book. Also included on this last page will be an explanation relating the biological classification to a real world situation such as organizing music.

OBJECTIVES

Science – Biology

- The student knows that taxonomy is a branching classification based on the shared characteristics of organisms and can change as new discoveries are made.
- The student will define taxonomy and recognize the importance of a standardized taxonomic system to the scientific community.
- The student will categorize organisms using a hierarchical classification system based on similarities and differences shared among groups.
- The student will compare characteristics of taxonomic groups, including archaea, bacteria, protists, fungi, plants, and animals.

Mathematics

- The student applies the concepts of similarity to justify properties of figures and solve problems.
- The student will describe the effect on perimeter, area, and volume when one or more dimensions of a figure are changed and apply this idea in solving problems.
- The student will compare and analyze various methods for solving a real-life problem.
- The student will communicate methods used, analyses conducted, and conclusions drawn for a data-analysis project by written report, visual display, oral report, or multi-media presentation.

Language Arts

- The student writes in a variety of forms, including business, personal, literary, and persuasive texts, for various audiences and purposes.
- The student will organize ideas in writing to ensure coherence, logical progression, and support for ideas.
- The student will write to report and describe using effective word choice, structure, and sentence forms with emphasis on organizing logical arguments with clearly related definitions, theses, and evidence.

R. M. Capraro, M. M. Capraro, J. Morgan and J. Scheurich, (eds.), A Companion to Interdisciplinary STEM Project-Based Learning: For Teachers by Teachers, 49–53.

MATERIALS NEEDED

- White sheets of paper
- Plastic cover or folder
- Internet access
- Biology book
- Glue
- Scissors
- Printer

STUDENT INTRODUCTION

James has a huge biology exam tomorrow, but he has not looked over any of his notes. He knows the exam covers the six kingdoms of life and the characteristics of each kingdom; however, he does not have enough time to read the entire chapter before his exam tomorrow. If James does not do well, he will not be able to participate in extracurricular activities. To help James prepare for this exam, relate biological classification to a real-world classifying activity such as organizing music to help him understand the importance of taxonomy. You will create a book explaining the similarities of classification techniques and describe the six kingdoms of life, their characteristics, and examples of organisms from each kingdom. The information may be organized in any way, but the book must look very neat and professional and be easy to read and understand.

DAY 1

Engagement

To initiate the project, the teacher may begin by having an open discussion about the biological world. This discussion should include how and why the world is classified. The teacher should create a dialogue with students and have them discuss how and why the world is grouped together. Then the teacher can present the material in the Student Introduction section above. Students should connect classification in science to their real lives, so the teacher could have students brainstorm to identify these connections.

Special consideration. Some students might not know all of the kingdoms. They likely will be able to name the more well known, such as the animal and plant kingdoms, but they might not have a firm grasp on the newer and less talked-about kingdoms.

DAY 2

Explanation

The teacher needs to go over the requirements for the book, which must include the following:
- A title that describes the contents of the book.
- Personal identification information on the cover page of the book.
- A page for each of the six kingdoms. Each page should contain the name of the kingdom in English and the important characteristics of organisms in that kingdom (e.g., cell type; cell structures; body type, i.e. uni- or multi-cellular; nutrition type; and motility). Also, something special must be mentioned about each kingdom. Finally, the Latin names and images (photos or drawings) of one or more organisms from each kingdom should be included.
- A word bank that contains the correct definition of all scientific vocabulary used, such as autotroph, heterotroph, etc and real life connection example. This should be on a separate sheet of paper.
- A bibliography that contains any books or web pages that were used to obtain images. Students may print out photographs from the web, but they must list the website and its title in the bibliography. Each bibliography will need to include at least three sources. If students do not cite where they found the information, they will not receive credit. If the teacher has access to an interactive whiteboard, this requirement provides a good opportunity to utilize this technology to demonstrate how to conduct an Internet search for information about the different kingdoms.
- Other requirements: Handwriting must be legible (if the teacher cannot read the students' work, no credit will be awarded), and the book must be in a plastic cover.

DAYS 3 AND 4

Exploration

During the next class period, present the student handout and have students use their biology book to do basic research. At this time, the teacher should divide students into groups and have the groups divide up the work so that each student will be working on one or two kingdoms. The students will spend the majority of this class period researching the six kingdoms to determine what information to use or leave out. It is a good idea to either have computers and books in the room or reserve a suitable location ahead of time.

DAY 5

Evaluation

Formative Assessment 1. During this class period, the students will put their books together. However, before the books are constructed, the teacher will collect the pages and pass out a different book to each group; no group should review its own book. The groups will then read over the books for content and grade them using the rubric provided by the teacher. The students will use the book content rubric (see Appendix A) to evaluate the book they are given. After the students review their assigned book, the teacher should review the scores to assess if any of the groups' books are severely lacking before the summative assessment. If a group needs to revise its content, the teacher should return the book to the appropriate group and have the students make corrections.

Formative Assessment 2. Once all of the books have the appropriate content, the books should be assembled and redistributed to different reviewers. This time, the reviewers evaluate the aesthetics of the book. The groups should use the book aesthetics rubric in the appendix (see Appendix B).

Summative Assessment. The teacher should use the student evaluations in the content and aesthetics rubrics as well as the teacher grading rubric (see Appendix C) to evaluate the groups' projects and performance overall.

Extension

Students will extend their book to incorporate the feedback received in the formative assessment and adding a two page executive summary.

Rubrics that were used in this lesson:

APPENDIX A

Kingdom Book Content Rubric

BOOK CONTENT	*Proficient*	*Meets*	*Making Progress*	*Score*
Cell Type	3 Contains necessary cell type information with explanations.	2 Contains adequate cell type information but is not thorough.	1 or 0 Missing pertinent cell type information.	
Body Form	3 Contains necessary body form information with explanations.	2 Contains adequate body form information but is not thorough.	1 or 0 Missing pertinent body form information.	
Cell Wall Composition	3 Contains necessary cell wall information with composition if needed.	2 Contains adequate cell wall information but is not thorough.	1 or 0 Missing pertinent cell wall information.	
Movement	5 or 4 Contains detailed information regarding movement.	3 or 2 Contains adequate information regarding movement.	1 or 0 Missing pertinent movement information.	
Reproduction Methodology	5 or 4 Contains a thorough explanation of reproduction methods.	3 or 2 Contains adequate information regarding reproduction methods.	1 or 0 Missing pertinent reproduction methods information.	
Evolution	5 or 4 Contains a thorough explanation of evolutionary significance.	3 or 2 Contains adequate information regarding evolutionary significance.	1 or 0 Missing pertinent information about evolutionary significance.	
Miscellaneous Information	5 or 4 Contains four or five additional miscellaneous facts.	3 or 2 Contains two or three additional miscellaneous facts.	1 or 0 Contains one or fewer additional miscellaneous facts.	
Case Studies	5 or 4 Contains six pages representing each of the six kingdoms with all required information.	3 or 2 Contains 3 pages representing 3 kingdoms with some of the information.	1 or 0 Contain 1 page with one kingdom with minimal information.	

APPENDIX B

Formative Assessment 2 – Book Aesthetics Rubric

Book Aesthetics			
Presentation	3 Highly creative and original.	2 Some creativity; missing one of the required elements.	1 or 0 Missing two or more of the required elements.
Visual Artistry	5 or 4 Highly creative and original. Contains appropriate illustrations.	3 or 2 Some creativity; contains illustrations that are less appropriate.	1 or 0 Little creativity or originality. Lacking adequate illustrations.
Neatness	5 or 4 Neat, organized, and appealing.	3 or 2 Not as neat or organized. Some errors are present.	1 or 0 Carelessly prepared; sloppy. Many errors.

APPENDIX C

Teacher Grading Rubric

Teacher Grading Rubric	
For this score...	Your book should look like this:
100–90	The book shows exceptional student understanding of material and is very polished and professional in appearance. All six kingdoms are included; characteristics of each kingdom are described completely and accurately; the name and image of at least one organism from each kingdom are provided; there are few spelling and grammatical errors; the book is neat; and the book includes a heading, word bank, real world connection, and appropriate title.
89 –80	All six kingdoms are included; characteristics of each kingdom are described completely and accurately; the name and image of at least one organism from each kingdom are provided; there are few spelling and grammatical errors; the book is neat; and the book includes a heading, word bank, real world connection, and appropriate title.
79 –70	All six kingdoms are included, some characteristics of the kingdoms may be left out or incorrectly described, the name and image of at least one organism from each kingdom are provided, the book is a little sloppy, there are several spelling and grammatical errors, and the book is missing one of the following: heading, word bank, real world connection, or title.
69 and below	The book is missing one or more kingdoms, there are many errors in the descriptions of the characteristics of the kingdoms, the book is very sloppy, there are many spelling and grammatical errors, and the book is missing more than one of the following: heading, word bank, real world connection, or title.

Lee Whitehead
Biology Teacher,
Emmett J. Conrad High School, Dallas ISD

M. SENCER CORLU

7. ARCTIC SHRINKAGE

WELL DEFINED OUTCOME

The student will create a persuasive speech to convince the American public that the Arctic region is shrinking compared to the past.

INTRODUCTION

Arctic shrinkage is the decrease in size of the Arctic region. In this project, students will use different geometrical methods to prove that there has been a significant decrease in the Arctic area over the last 30 years. Students will first investigate the environmental issues regarding global warming and discover the meaning of some of the most important terms used in environmental science. After this introduction, they will explore the basic features of Google Earth, free software that will help them with their investigation. Although area calculations of regular and irregular shapes are the focused content area of this activity, students will also have the opportunity to learn about the mathematics of making calculations on a spherical surface rather than a plane.

OBJECTIVES

Mathematics

- Students will think geometrically and use spatial reasoning to understand mathematical concepts and relationships.
- Students will recognize the relationship between algebra and geometry. Students will understand the connection between algebra and geometry and use the tools of one to solve problems in the other.
- Students will perceive the connection between geometry, other mathematics, and other disciplines. Students will use geometric ideas, relationships, and properties to solve problems.
- Students will be given tools for geometric thinking. Students will use a variety of representations (concrete, pictorial, numerical, symbolic, graphical, and verbal), tools, and technology (including calculators with graphing capabilities, data collection devices, and computers) to solve problems and to represent and transform figures and analyze relationships.

Science

- Students should understand that certain types of questions can be answered by investigations and that methods, models, and conclusions built from these investigations change as new observations are made.
- Students will learn that environments change, and they will recognize the relationship between carrying capacity and changes in populations and ecosystems.
- Students will describe how environmental change can impact ecosystem stability.
- Students will organize, analyze, evaluate, build models, make inferences, and predict trends from data.
- Students will perform calculations using dimensional analysis, significant digits, and scientific notation.

Geography

- Students will examine people, places, and environments at local, regional, national, and international scales from the spatial and ecological perspectives of geography.
- Students will describe the influence of geography on events of the past and present.

R. M. Capraro, M. M. Capraro, J. Morgan and J. Scheurich, (eds.), A Companion to Interdisciplinary STEM Project-Based Learning: For Teachers by Teachers, 55–64.

- Students will identify the physical processes that shape patterns in the physical environment; the characteristics of major land forms, climates, and ecosystems and their interrelationships; the political, economic, and social processes that shape cultural patterns of regions; types and patterns of settlement; the distribution and movement of world population; relationships among people, places, and environments; and the concept of region.

MATERIALS NEEDED

- The list of relevant vocabulary (some suggestions are provided in *Exploration* section)
- Colored pencils
- Poster boards in various colors
- Google Earth, free software that can be downloaded from the Internet
- One computer per student to maximize the effectiveness of this activity
- An LCD projector to show the videos
- A graphical calculator for the extension activity

STUDENT INTRODUCTION

Arctic shrinkage is the decrease in size of the Arctic region, which is caused by a change in the regional climate as a result of global warming. Recent projections of sea ice loss suggest that the Arctic Ocean will likely be free of summer sea ice sometime between 2059 and 2078. Because of the rapid response of the Arctic to global warming, it is often seen as a high-sensitivity indicator of climate change. Changes in temperature and precipitation patterns increase the frequency, duration, and intensity of extreme weather events, such as floods, droughts, heat waves, and tornadoes. Other effects of global warming include higher or lower agricultural yields, further glacial retreat, reduced summer stream flows, and species extinction. As a further effect of global warming, diseases like malaria have returned to areas where they had previously been eliminated.

As a young researcher assigned to prove there has been significant shrinkage of the sea ice area of the Arctic region, you are given satellite images of the region taken during the last 20 years. You must show the public that the rate at which the Arctic region is shrinking has increased over the years. To do this, you must learn how to use the software Google Earth and refresh your knowledge of area calculations. Do not underestimate these tasks, because the Arctic region is not composed of regular geometrical shapes, such as squares, rectangles or triangles.

Good luck!

ENGAGEMENT

The teacher starts off the project by introducing a KWL (What do I know? – What do I want to know? – What did I learn) chart of "Global Warming." Students will complete the K and W columns in their journals. After they complete this, the teacher will open a discussion about the possible causes of global warming. Students will watch an interactive video on global warming at http://epa.gov/climatechange/kids/global_warming_version2.html and compare the information from the video to the ideas they came up with during the discussion.

Two other video clips that can be shown to capture student interest are http://www.youtube.com/watch?v=wA70HAvUYpc (an approximately seven-minute long video including footage of some beautiful scenes of our Earth as seen from space) and http://www.youtube.com/watch?v=mGZbiWy47bA (a minute-long clip on Arctic shrinkage). This part of the activity will be completed in about one class period.

EXPLORATION

The next time the teacher meets with the students, he/she will present the story in the student introduction section above. The teacher will then divide the class into groups and assign an equal number of environmental terms to each group of students. Students will investigate the meaning of each term, individually record definitions in their journals, display an explanation of each term visually on a poster board as a team, and present the definitions to the rest of the class. The following terms can be used for this investigation: global warming, greenhouse effect, greenhouse gases, Arctic shrinkage, Kyoto protocol, ozone layer, greenhouse emission, carbon cycle, Earth's atmosphere, fossil fuels, polar city, sea ice, nautical mile as it pertains to measurement.

Students will need to learn not only the terms assigned to them but also be prepared to ask questions during the other groups' presentations. They will have some constraints to their exploration, which are:

- They have to use at least two resources for each vocabulary term and should use their own words to record the meaning of each term in their journals.
- They have to draw one large illustration for each term to aid their explanation to the other groups.

The teacher will observe the students working in teams and evaluate their progress. Students will evaluate the effectiveness of the other teams' presentations. The teacher will distribute a modified version of Rubric 1 (see Appendix A) so that each group can evaluate their peers' presentations. This part of the project will take about two class periods.

EXPLANATION

The teacher will assign each group to one computer and demonstrate the different features of Google Earth. To prepare for this activity, the teacher should either help students follow the online instructions for downloading and installing Google Earth software at http://earth.google.com/ or install the program in advance. After installing the software, students can open the program by clicking on the icon on the desktop.

The first part of this phase, which should be covered in two periods, teaches students to use Google Earth and discover the shortest distance between any two cities in the U.S. However, students do not know whether this distance is shortest for a person walking, driving, or flying. They are not given the unit of the distance, either. Students are expected to choose the most appropriate measurement unit in their calculations; miles, kilometers, or nautical miles. Thus, students are asked to give a detailed calculation of their findings in a chart and answer the following questions:

- If you could draw a straight line between College Station and Waco, would it really be straight? Explain.
- Why do sailors/aviators prefer using nautical miles rather than ordinary miles to calculate the distance between two distinct points on the map? Which is bigger: one nautical mile or one mile?
- Why are roads not built straight? What geological features may cause this?
- Rubric 2 (see Appendix B) will be used to assess student progress and is given at the end of this chapter.

Specific explanations

On the left pane of Google Earth, the "Directions" tab can be used to calculate the shortest distance by car between any two places on Earth (see Figure 1).

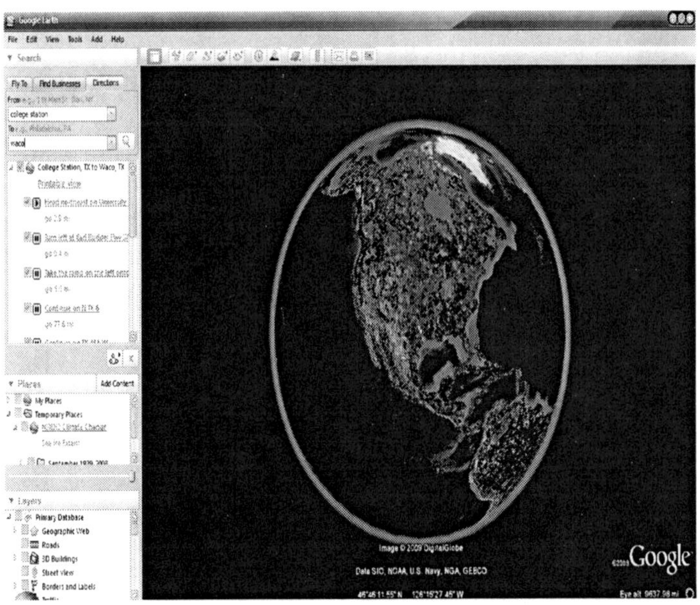

Figure 1. Google Earth opening page.

Alternatively, the ruler symbol can be clicked to calculate the direct distance between any two places on Earth. Google Earth can also be asked to calculate the distance of a drawn path as in Figure 2.

The final part of the activity is for more advanced students who are familiar with trigonometry; this section includes techniques for calculating irregular shapes. The teacher introduces the challenge of students convincing the U.S. government to sign the Kyoto protocol by proving the shrinkage in the Arctic region. In order to do this, students should learn the advanced features of Google Earth. In the top part of the toolbar, there is a polygon feature that can be used to draw a shape that covers a region of choice in order to calculate the area (see Figure 3).

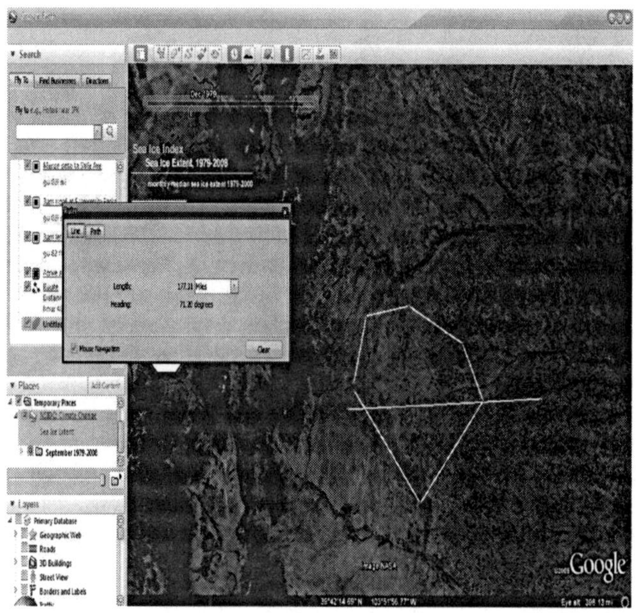

Figure 2. Ruler feature and distance calculation in Google Earth.

Figure 3. Polygon feature in Google Earth.

Selecting the clock icon in the top part of the toolbar shows the look of any place on earth at different times in history. See a view of College Station in 2004 in Figure 4.

After learning to navigate the advanced features of Google Earth, students should download and run Arctic Sea Ice extension (1979–2008) available at http://nsidc.org/data/virtual_globes/seaice/NSIDC_SepSeaIce Extent.kmz. This extension will let students see the change in the Arctic Sea from 1979 to 2008 by using the history tool (the clock symbol in the top toolbar). See Figure 5.

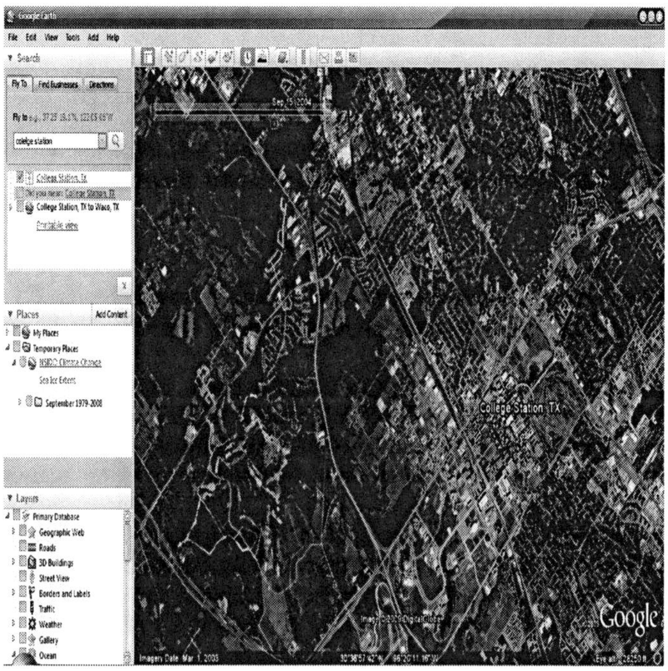

Figure 4. History tool in Google Earth.

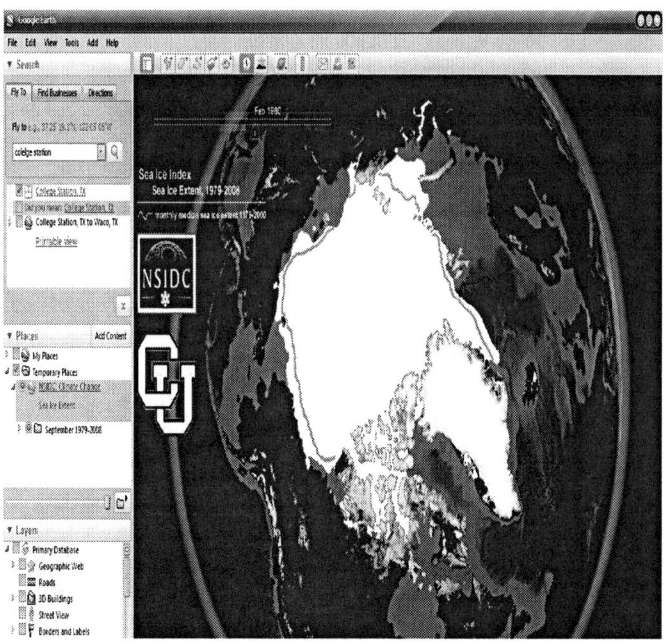

Figure 5. Google Earth after the Arctic Sea Ice Extend add-on was installed.

Students will measure angles of triangles with protractors and calculate the area of triangles using the "Law of Sines," "Heron's Formula," and "Divide and Conquer." Students will also use the "Monte Carlo" procedure to predict area via probability as an extension. Students will analyze solutions from the different procedures for calculating area to decide the accuracy of their results. Equations used in this activity are:

$$Law\ of\ Sines: \frac{sin(\angle A)}{a} = \frac{sin(\angle B)}{b} = \frac{sin(\angle C)}{c} \quad (1)$$

$$Heron's\ Area\ Formula: \sqrt{s(s-a)(s-b)(s-c)}\ \ where\ s = (a+b+)/2 \quad (2)$$

Once students have completed their research and calculations, they will formulate the learned information into a persuasive speech to convince Americans of the problem of Arctic shrinkage.

EXTENSION

The "Monte Carlo" method can be used as an extension. Rubric 3 (see Appendix C) is a tool for assessing student performance. This section of the project-based learning activity should take approximately two periods.

Students will use probability to find the area of an irregular shape. Supposing that a rectangle holding the irregular shape was a dartboard and darts were randomly thrown at the dartboard, the percentage of darts that hit the irregular object compared to the total number of darts thrown should approximate the area of the irregular object. This calculation can be accomplished by using a graphing calculator's random number operations.

The grid used in the example problem below is 18 by 10 units (see Figure 6). Therefore, consider the coordinates of the dartboard 18 by 10. Use a calculator to chart 20 dart tosses and tabulate how many darts hit within the interior of the irregular shape. Decide whether to include the boundary points between the two shapes. Try the experiment using both. Place the random *x* values in L1 of your calculator and the random *y* values in L2 of your calculator:

MATH – PRB – (5) randInt()

Enter: randInt(0,18,20) –> L1

Enter: randInt(0,10,20) –> L2

Make a chart of L1, L2 values and mark all hits within the interior with a symbol (+).

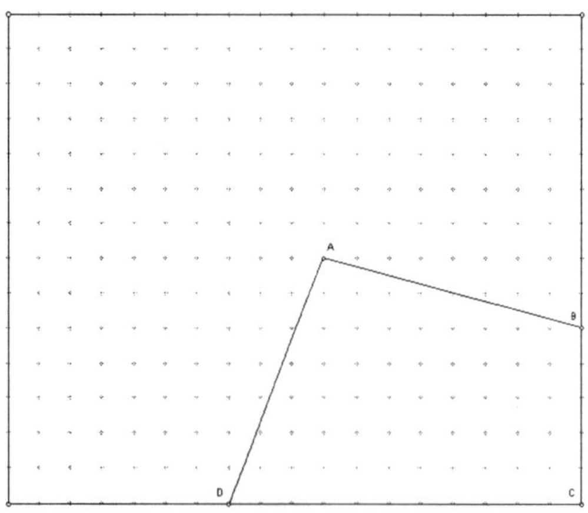

Figure 6. Probability grid.

What percent of the time did the coordinates end up within the interior of the irregular shape? In this example, if the interior region was hit 6 times out of 20 or 30, then the total area of the rectangle is 0.30*(180) = 54 square units (see Figure 7).

The teacher asks the students how they could improve the accuracy of this method and whether including the boundary makes the answer more accurate or less accurate.

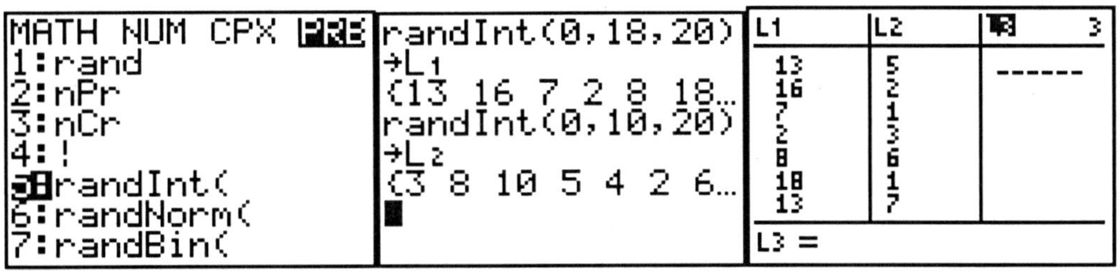

Figure 7. TI-84 calculator screens.

Another extension of this activity could be to create multimedia presentations to further convince citizens of the need to correct the Arctic shrinkage. Videos, commercials, or websites could be developed to complete this task.

APPENDIX A

Student Research and Vocabulary

	RESEARCH (by teacher)	*PRESENTATION (by other students)*	*GROUP WORK (by teacher)*
4–3	Complete understanding of each term, and students' notes in their journals are their own words rather than a copy-paste from the resource. A variety of resources are used, and they are indicated separately.	Great visualization and a very clear presentation of each assigned term.	Each student worked diligently, and the group shared responsibilities according to each member's strength.
2–1	Some terms are misunderstood or the explanations are not clear. Cited resources are not indicated clearly.	Some of the terms are not presented at the required level, or poor visualization and not very clear explanations of the terms.	Some students did most of the work, while others made a minimal contribution.
0	No attempt or mostly copy-paste work.	No attempt or totally careless presentation.	All the work was done by certain students, others made no contribution.
Comments			

APPENDIX B

Using Google Earth

	Understanding the problem	*Planning a solution*	*Getting an answer*
4–3	Complete understanding of the problem; students referred to the Earth's round shape.	Students successfully used the ruler feature of GoogleEarth or graph paper and their plan could have led to a correct solution if implemented properly.	Correct answer and correct label for the answer.
2–1	Part of the problem was misunderstood or misinterpreted.	Partially correct plan based on part of the problem being interpreted correctly.	Copying error, computational error, or partial answer for a problem with multiple answers.
0	Complete misunderstanding of the problem.	No attempt or totally inappropriate plan.	No answer or wrong answer based on an inappropriate plan.

APPENDIX C

Rubric 3 (Extension activity)

	Understanding the problem
4	Students accurately used all methods for calculating the area of the irregular shape and showed all work and answers. Students compared results and made statements regarding the accuracy of their answers.
3	Students used all methods for calculating the area of the irregular shape and showed all work and answers. Students had a minor error in computation on one or two of the methods.
2	Students used all methods for calculating the area of the irregular shape and showed all work and answers. Students had errors in computation on more than two of the methods.
1	Students did not use all area methods and/or made more than five errors in computation.
0	Virtually no effort was placed on calculating area of the irregular shape using various methods.

APPENDIX D

Summative Assessment

	Understanding the problem	*Planning a solution*	*Getting an answer*
4–3	Complete understanding of the problem. Students used both ratio and difference methods of comparing two quantities and gave a valid explanation of the Arctic shrinkage issue.	Students used a plan to calculate the area and perimeter of the Arctic region, a number of methods as explained in the extension section, the history tool to see and calculate area and perimeter in the past with meaningful intervals, and provided persuasive solutions to the problem.	Correct answer, correctly labled, and good representation in a chart or table to show the change in the area and perimeter of the Arctic region. Students referred to the fact that their calculations were within the error caused by the shape of the Earth.
2–1	Part of the problem was misunderstood or misinterpreted. Students were not aware of the limitations of their calculations. Shape of the earth is not mentioned. Explanation of the issue is skewed, but can be argued correctly.	Partially correct plan based on part of the problem being interpreted correctly. Students couldn't show the intensity of shrinkage over the years either by not including the perimeter or the area. Students provided a weak solution to the problem that was not particularly persuasive.	Copying error, computational error, or partial answer for a problem with multiple answers.
0–1	Complete misunderstanding of the problem with little explanation of the issue	No attempt or totally inappropriate plan; students were unable to come up with a method to estimate the area and perimeter of the Arctic region, neither in the past nor at present. Students provided no real solution to the problem.	No answer or wrong answer based on an inappropriate plan.

REFERENCE

Yohe, J. (2002). Retrieved June 10, 2009, from Area of Irregular Shapes: http://www.susq-town.org/ yohe/IrregularArea/index.htm

M. Sencer Corlu
Aggie STEM Center,
Texas A&M University

STRUCTURES

CHARLES L. CREGG

8. TECHNOLOGY SYSTEMS (BUILDING BALSA WOOD BRIDGES)

WELL-DEFINED OUTCOME

Students will understand why applying problem solving techniques and the civil engineering design process is so important for developing new bridge designs in the 21st century.

INTRODUCTION/BACKGROUND

Civil engineering is the branch of engineering that specializes in the design and construction of structures such as bridges, roads, and dams. In this project, students will study the engineering design process in the application of math, science, and technology to create a system, component, or process to meet a human need or solve a problem in real world time. The seven steps of the engineering design process are: identify the problem, define the problem, develop alternative solutions, analyze and compare alternative solutions, select the best alternative, implement the solution, and evaluate the results.

OBJECTIVES

Technology

- Students will know how to use computers, software, and printers, which play an important role in the engineering design process.

Mathematics

- Students will be able to use mathematical models to represent and understand quantitative relationships.
- Students will use numerical computations to judge reasonable estimates and results.
- Students will understand the measurement attributes of bridge design in areas of units, systems, scales, and the processing of measurements.
- Students will calculate cost and estimate cost factors in the areas of site cost (excavation, deck, and support) and truss cost (material, connection, and product).

Science

- Students will be able to identify questions and concepts that will lead them when conducting scientific investigations.
- Students will design and conduct scientific investigations that will improve communication skills, help them to formulate and revise models, and use logic and evidence in deductive reasoning.

Physics

- Students will gain understanding and a working knowledge of compression, tension, span, and load.
- Students will be able to describe what a truss is as well as several types of trusses.
- Students will gain understanding and a working knowledge of the differences between truss bridges, beam bridges, arches, suspension bridges, and cable-stayed bridges.

R. M. Capraro, M. M. Capraro, J. Morgan and J. Scheurich, (eds.), A Companion to Interdisciplinary STEM Project-Based Learning: For Teachers by Teachers, 67–77.

English

- Students will use proper grammar, spelling, and punctuation to document hypotheses, to formulate questions that address these hypotheses, and to make observations about data (collection, evaluation, and comparison).

Safety precautions

- Students will understand and comply with general lab safety rules at all times.
- Students will wear safety eye wear at all times.
- Proper use of all tools and sharp knives/blades will be enforced.
- Students will be monitored closely in the lab.

MATERIALS NEEDED

Bridge construction tools and materials:
- Computer
- WPBD 2007 software
- Copier paper
- Cardboard
- Wax paper
- T-Pins
- Masking tape
- Scotch tape
- Glue
- Sandpaper
- Timber cutter
- Ruler (steel, wood or plastic)
- Protractor
- Structural tester

STUDENT INTRODUCTION

Students will design a bridge structure using software developed at West Point Military Academy. The website http://bridgecontest.usma.edu introduces real-world time, civil engineering design processes to students. The students' bridge designs must support a certain weight load within a specified cost constraint.

Students will test their bridge design for stability using a test simulator provided in the software program. Once the bridge design passes the simulator test, designs will be approved for printing. Structural truss member joints, a cost calculation, and a material list will also need to be printed out.

Students will transfer their design to the balsa wood, cut balsa wood, and construct the bridge structure. Then students will test the final structure for maximum load-bearing load results.

ENGAGEMENT

The teacher will discuss and demonstrate how to use the West Point Bridge Designer tutorial to students. The West Point Bridge Designer 2007 allows users to experience the engineering design process in simplified form. The program allows its users to design a highway bridge in much the same way that practicing civil engineers design real highways bridges.

The program will present students with the requirement of designing a steel truss bridge that carries a two-lane highway across a river. Students may choose from a wide variety of site configurations for their bridge. Each will cause the bridge to carry loads in different ways, and each has a different site cost. Students can develop a design for their bridge by drawing a picture of it on their computer screen.

Once a first design attempt is complete, the West Point Bridge Designer 2007 will test the bridge to see if it is strong enough to carry the specified highway loads. This test includes a full-color animation showing a truck crossing the bridge. If the design is strong enough, the truck will be able to cross the bridge successfully; if not, the structure will collapse. If a student's bridge collapses, he or she can strengthen it by changing the configuration of the bridge itself.

When the bridge can successfully carry the highway load without collapsing, students can continue to refine their design with the objective of minimizing cost while still ensuring that it is strong enough to carry the specified loads.

The West Point Bridge Designer 2007 will give students complete flexibility to create designs using any shape or configuration they want. Creating the design is fast and easy, so students will be able to experiment with many different alternative configurations as they work towards the best possible solution. The process used is similar to the process used by practicing civil engineers as they design real structures. In fact, the West Point Bridge Designer 2007 is similar to the computer-aided design (CAD) software used by practicing engineers, and it will help students in the same way CAD software helps them—by taking care of the heavy-duty mathematical calculations so that the program user can concentrate on the creative part of the design process.

EXPLORATION

Vocabulary for Bridge Designing/Building

Instructions. Student will use the internet to search for the following list of vocabulary words used in bridge designing/building. They can use www.dictionary. com or search engines such as Google, Yahoo, etc. Then they will use Microsoft Word to list the vocabulary word and definition. Students should include clip art related to bridges (at least three pictures). The instructor will print a hard copy of each student's work for them to put in their class binder.

- Angle
- Average
- Beam
- Bridge
- Compression
- Conclusion
- Constraint
- Criteria
- Critical load
- Design
- Dimension
- Efficiency
- End grain
- Force
- Hypothesis
- Inference
- Joint
- Lamination
- Load-bearing capacity
- Mass
- Model
- Modification
- Newton
- Racking
- Rigid
- Roadbed
- Scale
- Side grain
- Side view
- Sketch
- Span
- Substructure
- Tension
- Timber
- Truss

Exploration of Balsa Wood Properties

Balsa Wood. Balsa wood is a very interesting material that is typically used for the construction of model airplanes. Balsa is technically a hardwood because of the shape of the leaves. It is not the lightest wood available, but it is readily available, and has a very good strength to weight ratio. It is not considered to be a great material to use for the design of most things because its properties vary greatly.

As with most woods, balsa wood does not have well-defined mechanical properties. Its properties vary greatly with the grain direction and moisture in the wood, but in order to do design calculations, some rough properties are needed. Below is a table of balsa wood properties for conducting design calculations. Other information can be found in the Appendix A of this chapter, in reference books, or online.

Table 1. Balsa wood properties

Properties

Light under 5 lbs/ft^3 (.00289lbs/in^3; 1.31 g/in^3)
Medium 5 to 9 lbs^ft 3 (.00289 - .00521 lbs/in^3; 1.31g/in^3 - 2.36 g/in^3)
Heavy over 9 lbs/ft^3 (0.00521 lbs/in^3; 2.36 g/in^3)

Compressive strength

Light	680 lbs/in^2
Medium	1750 lbs/in^2
Heavy	2830 lbs/in^2

Tensile Strength

Light	1100 lbs/in^2
Medium	2890 lbs/in^2
Heavy	4670 lbs/in^2

Elastic Modulus in compression

All 66700+/-10300 lbs/in^2

Elastic Modulus in Tension

All 185,300 +/- 65400 lbs/in^2

EXPLANATION

Types of Bridges and Bridge Connections

The students will be researching and experimenting with the different types of bridges listed below during the design and layout process. The different types of bridges are:

Beam bridge. The beam bridge was derived from the log bridge. It is built from shallow steel beams, box girders, and concrete. Highway overpasses, flyovers, and walkways are often beam bridges. A horizontal beam supported at the beam bridge's ends comprises its structure. The construction of a beam bridge is the simplest of all the types of bridges.

Truss bridge. A truss bridge is built by connecting straight elements with the help of pin joints. In the past, truss bridges used timbers for compression members due to the abundance of wood in the United States and iron rods for tension members. Truss bridges were commonly constructed from the 1870s to the 1930s. The deck truss railroad bridge that extends over the Erie Canal is one of the many famous truss bridges.

Arch bridge. This bridge earns its name from its arch shape, and it has supports at both ends. The weight of an arch-shaped bridge is forced into the supports at either end. The Mycenaean Arkadiko Bridge in Greece of 1300 BC is the oldest existing arch-shaped bridge. Etruscans and the ancient Greeks used arches in construction, but the Romans were foremost in popularizing the use of arches in the construction of bridges. Arch bridges have now evolved into compression arch suspended-deck bridges, which enable the use of strongly tensile and light materials in their construction.

Suspension bridge. Bridges in this category are suspended from cables. The suspension cables are anchored at each end of the bridge. The load on the bridge is transmitted into the tension in the cables, and from the cables into the towers and into the cable anchors on each shore. The Golden Gate Bridge in the United States, the Chinese Tsing Ma Bridge, and the English Humber Bridge are some famous suspension bridges.

Cable-stayed bridge. Structured like suspension bridges, cable-stayed bridges differ in the amount of cable used. Less cable is required, and consequently, the towers holding the cables are shorter. Two variants of cable-stayed bridges exist. In the harp design, cables are attached to multiple points of the tower thus making them parallel. In the fan variant of design, all the cables connect to the tower or pass over it. Cable Bridge boasts of being the first cable-stayed bridge in the United States. Centennial Bridge is another well-known, cable-stayed bridge.

Cantilever bridge. Cantilevers are the structures that project along the x-axis in space. They are supported only on one end. Bridges intended to carry lesser traffic may use simple beams whereas those aimed at handling larger traffic make use of trusses or box girders. The 1800-foot long Canadian Quebec Bridge and the 1400-foot San Francisco-Oakland Bay Bridge are two examples of cantilever bridges.

EVALUATION

Rubrics

A rubric is one of the best tools available to clarify expectations for a project. It is best to distribute a rubric to the students (such as the one in Appendix B) when the project is assigned. Also, the rubric will make the final project reports and presentations easier to grade. One rubric may include all features of the project's grade or different rubrics may be used for different portions of the grade (i.e., presentation, report, performance, etc.).

Final Load Testing of the Structure

The steps in the process of a load test are as follows:

Under construction. The balsa wood model bridge is not yet completed, or it has been changed since the last load test.

Unsafe. The structural model has been load test, and one or more members are not strong enough to safely carry the specified loads.

Safe. The structural model has been load tested, and all members are strong enough to safely carry the specified loads.

Students will test the final structure for maximum load-bearing-load results. This means that the bridge model will be tested to failure. Students will compare the performance of the actual bridge with the results of the simulation in the West Point Bridge Designer 2007 software.

Report Containing the "Cost" of the Bridge

The West Point Bridge Designer 2007 software will automatically calculate the cost of the student's bridge as they design it. The cost changes as the design is updated.

There are two main components in figuring the total cost:

- Site cost consists of three areas (excavation cost, deck cost, and support cost).
- Truss cost consists of three areas (material cost, connection cost, and product cost).

In minimizing the total cost of a bridge, students will find that they cannot minimize the following cost areas:

- Site cost
- Material cost
- Connection cost
- Product cost

Prediction of Failure and Comparisons of the Actual Failed Bridge Design

Design Process. The following nine design process steps are listed only as a sample for predicting simulation failure and comparing actual failed bridge designs:

- Failure mechanisms
- Material selection
- Cross section selection and analysis
- Incorporating pultrusions
- Analytical analyses/analytical optimization
- ANSYS analysis of baseline design
- ANSYS validation with analytical predictions
- ANSYS model of optimized design
- Manufacturing/prototype/testing

EXTENSION

The engineering design process included an iterative improvement process. Students can take their bridge design to the next level by either increasing the maximum load to be carried or by decreasing the cost of a bridge to carry the same maximum load. Evaluation for the new improved bridge can be tied to the percent increase in load, percent decrease in cost, or a combination of these two factors.

APPENDIX A

Strength of Balsa Wood Compared to Other Woods

Species	*Weight Lbs./Cu. Ft.*	*Stiffness Strength*	*Bending Strength*	*Compression Strength*
Balsa	8	72	70	75
Balsa	10	100	100	100
Balsa	14	156	161	149
Spruce	28	230	260	289
Yellow Pine	28	222	277	288
Douglas Fir	30	241	291	341
Hickory	50	379	638	514
Oak	48	295	430	366
Basswood	26	261	288	288
Black Walnut	37	301	506	512

Balsa Sticks, Weight in Ounces

SIZE 36" x		*Number of pieces*	*STOCK DENSITY in pounds per cubic foot*					
			6	*8*	*10*	*12*	*14*	*16*
1/16"x	1/16"	16	.125	.167	.208	.250	.292	.333
	3/32"	16	.188	.250	.312	.375	.438	.500
	1/8"	8	.125	.167	.208	.250	.292	.333
	3/16"	8	.188	.250	.312	.375	.438	.500
	1/4"	4	.125	.167	.208	.250	.292	.333
	3/8"	4	.188	.250	.312	.375	.438	.500
	1/2"	2	.125	.167	.208	.250	.292	.333
3/32"x	3/32"	8	.141	.188	.234	.281	.326	.375
	1/8"	8	.188	.250	.312	.375	.438	.500
	3/16"	4	.141	.188	.234	.281	.326	.375
	1/4"	4	.188	.250	.312	.375	.438	.500
	3/8"	2	.141	.188	.234	.281	.326	.375
	1/2"	2	.188	.250	.312	.375	.438	.500
1/8"x	1/8"	4	.125	.167	.208	.250	.292	.333
	3/16"	4	.188	.250	.312	.375	.438	.500
	1/4"	4	.250	.333	.416	.500	.538	.667
	3/8"	2	.188	.250	.312	.375	.438	.500
	1/2"	1	.125	.167	.208	.250	.292	.333
3/16"x	3/16"	2	.141	.188	.234	.281	.326	.375
	1/4"	2	.188	.250	.312	.375	.438	.500
	3/8"	2	.281	.375	.469	.563	.656	.750
	1/2"	1	.188	.250	.312	.375	.438	.500
	3/4"	1	.281	.375	.469	.563	.656	.750
	1"	1	.375	.500	.625	.750	.876	1.00

Balsa Sticks, Weight in Ounces (continued)

1/4"x	1/4"	1	.125	.167	.208	.250	.292	.333
	3/8"	1	.188	.250	.312	.375	.438	.500
	1/2"	1	.250	.333	.416	.500	.583	.667
	3/4"	1	.375	.500	.625	.750	.876	1.00
	1"	1	.500	.667	.832	1.00	1.166	1.333
3/8"x	3/8"	1	.281	.375	.469	.563	.656	.750
	1/2"	1	.375	.500	.625	.750	.876	1.00
1/2"x	1/2"	1	.500	.667	.832	1.00	1.166	.1333
	1"	1	1.00	1.333	1.666	2.00	2.333	2.667
3/4"x	3/4"	1	1.125	1.500	1.875	2.250	2.625	3.00

APPENDIX B

Sample Rubric

CATEGORY	*4–3*	*2–1*	*0*
Understanding	Demonstrates complete understanding of the problem including the elements contributing to the solution process.	Demonstrates a limited understanding of the problem; ignores some elements of the solution.	Does not understand the problem or the solution.
Plan	Identifies and effectively follows the input, process, output, and feedback structure. Investigates the work of others, but develops an original, detailed plan. Selects and applies the most feasible plan among alternatives. Provides justifications to decisions such as cost, availability, etc. Shows proof of division of labor and responsibilities of group members. Members' roles are clearly defined, and how the team will function is described.	Misses some elements of the input, process, output, and feedback structure. Ignores the work of others or adapts a non-generic plan with minimal details. Does not consider alternative solutions, but applies a feasible plan. Some elements are mostly guess work rather than the result of extensive research and calculations.	Does not consider elements of the input, process, output, and feedback structure. Does not consider the work of others. Does not have a plan.
Design	Uses a variety of tools, equipment, machines, materials, and technology. Considers quality, reliability, safety, and maintenance issues of the design. Uses correct mathematical and/or scientific principles in the designs; all calculations are precise and appropriate, and technology is used. Applies the appropriate codes, laws, standards, or regulations, such as National Electrical Code (NEC), Occupational Safety and Health Administration (OSHA), American Society for Testing Materials (ASTM), standard symbols, and line weights.	Uses a limited number of tools, equipment, machines, materials, or technology. Fails to consider one or more of the quality, reliability, safety, and maintenance issues of the design. Uses mathematical and/or scientific principles with some errors including some calculation mistakes.	Does not use tools, equipment, machines, materials, or technology. Does not consider quality, reliability, safety, or maintenance issues of the design. Does not use mathematical or scientific principles.
Evaluation	Uses an assessment strategy to determine the risks and benefits of the activities and outputs. Describes how the design may affect individuals, societies, cultures, economies, and environments. Uses an engineering approach to problem-solving to improve the product. Applies critical-thinking strategies to the analysis and evaluation of solutions.	Describes the risks or benefits of the activities and outputs without a proper strategy. Uses a non-systematic approach to problem-solving to improve the product.	Does not know the risks or benefits of the activities. Does not have a strategy. Does not problem solve to improve the product.
Presentation	Uses written, verbal, and visual communication techniques consistent with industry standards; prepares a marketing plan for an idea, product, or service. Includes complete, correct, and neat sketches to describe system and/or machine elements.	Includes sketches that are not clear. Presentation of the plan or design has some missing elements and/or is not clear.	Does not have a presentation.

APPENDIX C

Examples of Student Work

Pictures of actual student built bridges are presented in figure 1 and figure 2 below. Additional information may be obtained by opening the examples of student bridge designs; including materials used, and cost factor information. To open the files, go to http://bridgecontest.usma.edu, download (it is free) the following items from the website onto your computer:

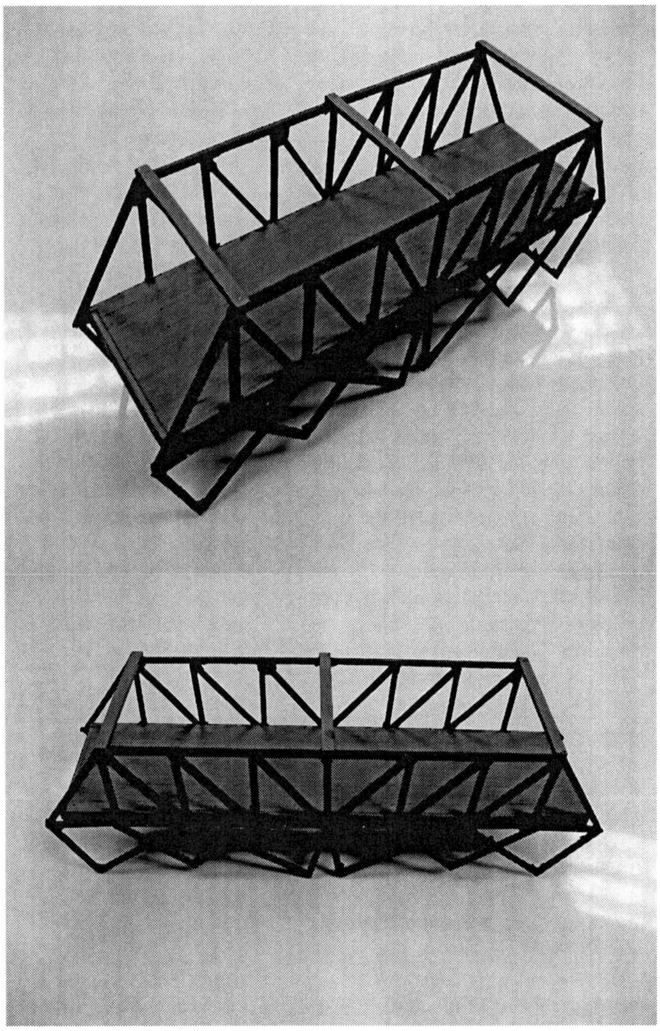

Figure 1. Example of simple student bridge designs.

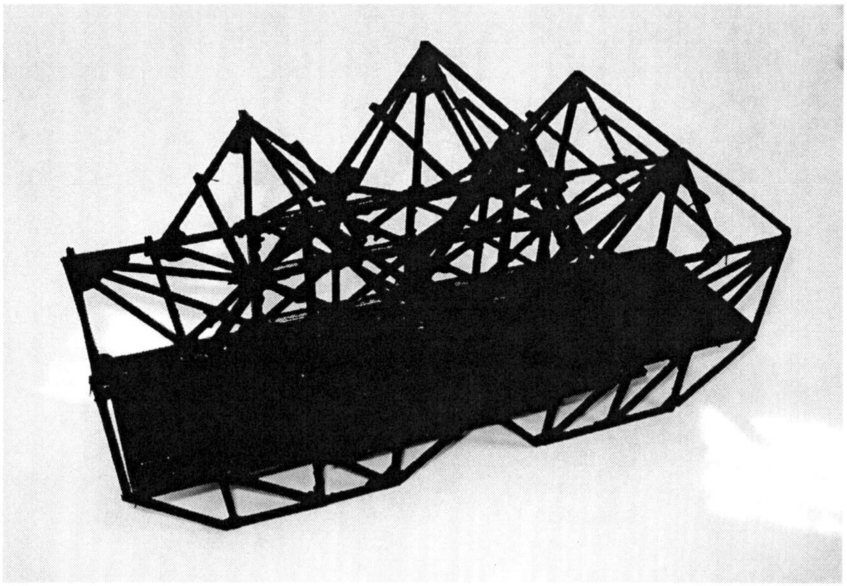

Figure 2. Example of a complex student bridge design.

Charles L. Cregg
Engineering Teacher,
Emmett J. Conrad High School, Dallas ISD

CRYSTAL J. WOODS AND ROBERT M. CAPRARO

9. CONRAD FOODS: PACKAGING "R" US

WELL-DEFINED OUTCOME

At the conclusion of this project, students will have designed and built packaging for their product that meets design specifications. Students will decide the shape and size of the packaging using measurement and geometry for their product. They will analyze various materials to determine the best choice for their product. Students will be able to use geometric concepts to determine the measurements of three-dimensional figures in order to solve problems connected to everyday experiences, in this case, product packaging. Based on the lateral and total surface area of the package, students will determine the total cost of the material they select.

INTRODUCTION/BACKGROUND

The purpose of this project is to create or engineer a container for a product out of material that is cost-efficient and effective for protecting a product. A common real world job is product packaging. The development and design of a product is only one part of being able to sell that product. The product has to be contained in something that protects it during shipment and from potentially, handling by machines. Additionally, being able to design a package that is cheap is important because if the package costs too much it will either decrease the profits made from the product or force the manufacturer to price the product too high which could cause customers to choose a competitor's cheaper product. This is one common job of an industrial distribution engineer. Students should understand and explore that every product they purchase is packaged for sale and that packaging can influence what a consumer purchases, complicated packaging may require special machines and that poor packaging results in broken unsellable products which creates losses for the manufacturer. For example, if the product designed is a soft drink, it has to be put into bottles or cans of some shape made of a material like aluminum, plastic, or glass. Each choice requires careful selection of how that item will be grouped for sale. Some may choose two-, three-, four-, or six- packs. Some may choose padded containers while others use cardboard. If the packaging breaks while in transit and the soft drink containers rupture the manufacturer will loose their investment in producing the product. At the start of the project, the teacher can have students talk about their experiences with various products and their packaging.

The teacher can use questions to prompt the discussion. Would the students prefer to have their favorite product packaged differently? Why are potato chips packaged in air filled bags? How else might potato chips be packaged? How are round objects packaged? Why are shoes packaged in rectangular prisms? Is there a better way to package toothpaste? What do you like or dislike about the product packaging? As students talk about the product packaging they like or dislike, they start to figure out what is important in packaging a particular product. After this discussion students should conduct research based on their discussion. This research engages them with the project and requires them to collect data in order to create and engineer the most efficient package for their products. The purpose of the research can include, as time allows, exploring possible product choices – or designing a new product, the job of industrial engineers, designing carriers versus bulk packaging, exploring how packaging is related to marketing, using recycled materials, or how packaging is related to how customers will use the product. They can also explore the various mathematics concepts they will need to develop their own packaging (e.g., three dimensional shapes, nets, and measurement and volume formulae.)

OBJECTIVES

Mathematics

When students complete this project, they will be able to:

- Find the lateral and total surface area of prisms, pyramids, and cylinders using concrete models and nets (two-dimensional models).
- Connect models of prisms, cylinders, pyramids, spheres, and cones to formulas for finding the volume of these objects.

R. M. Capraro, M. M. Capraro, J. Morgan and J. Scheurich, (eds.), A Companion to Interdisciplinary STEM Project-Based Learning: For Teachers by Teachers, 79–85.

- Estimate measurements and use formulas to solve application problems involving lateral and total surface area and volume.
- Identify and apply mathematics to everyday experiences, activities in and outside of school, other disciplines, and other mathematical topics.
- Use a problem-solving model that incorporates understanding the problem, making a plan, carrying out the plan, and evaluating the solution for reasonableness.
- Select or develop an appropriate problem-solving strategy from a variety of different types, including drawing a picture, looking for a pattern, systematic guessing and checking, acting it out, making a table, working a simpler problem, or working backwards to solve a problem.
- Select tools such as real objects, manipulatives, paper and pencil, and technology or techniques such as mental math, estimation, and number sense to solve problems.
- Communicate mathematical ideas using language; efficient tools; appropriate units; and graphical, numerical, physical, or algebraic mathematical models.
- Evaluate the effectiveness of different representations to communicate ideas.

Language Arts

When students complete this project, they will be able to:
- Provide an analysis of the relative value of specific data, facts, and ideas.
- Use graphics and illustrations to help explain concepts where appropriate.
- Use a variety of evaluative tools (e.g., self-made rubrics, peer reviews, teacher and expert evaluations) to examine the quality of the product.
- Speak clearly and effectively for a variety of purposes and audiences. Use the conventions of oral language effectively.
- Use informal, standard, and technical language effectively to meet the needs of purpose, audience, occasion, and task.
- Organize and present ideas and information according to the audience and purpose of the research.

MATERIALS NEEDED

- Scissors
- Colored pencils or markers
- Glue
- Formula chart (volume and area of three-dimensional shapes)
- Grid paper
- Yarn
- Ruler
- Computer (PowerPoint) or any presentation medium

STUDENT INTRODUCTION

Conrad Foods is a food company in town. In today's economy, Conrad Foods is taking active steps to achieve cost-efficient production. One particular area where Conrad Foods decided to save money is in packaging their food products. However, the company does not want to compromise the attractiveness and efficiency of their packaging. Because teenagers constitute a large percentage of Conrad Foods' consumers, the CEO, Mr. Brown, has decided to get ideas from you on the development of the company's new packaging. Thus, Mr. Brown would like each of you to develop a cost-effective packaging for a newly developed food product. After you have determined a product you will design and create your packaging, then you are going to present your product and its packaging in both written and oral forms including a packaging cost and the most efficient packaging is preferred. The packaging will be judged based on cost-efficiency, durability, protection, and attractiveness.

ENGAGEMENT

Day 1

The teacher starts off the project with an open discussion about what motivates the students to buy certain food products. What role does packaging play in their decision to buy food products? During this discussion, the teacher stimulates students' interest in developing an attractive, durable, secure, and cost-efficient packaging for their food product. It is important to develop a common understanding for what is included in the terms attractive and effective packaging. While these seem like simple terms they are not and for student success a clear understanding here is essential. Students can write down or draw their favorite food packaging. Students can then discuss what makes their favorite food packaging attractive to them and if and how they could improve that particular packaging. The teacher draws a chart on the board, and students copy the chart into their journals. Students will engage in a teacher-led discussion to fill in the chart. An example is provided below in Table 1.

Table 1. Sample journal entry

Characteristics of attractive packaging	*Characteristics of effective packaging*
	Keeps the food product fresh Resistant to temperature Resistant to light Environmentally friendly Protects the product from breakage, spoilage, and theft.

Students should be encouraged to think about the cost of the packaging they suggest. The teacher may also display different packaged items (of assorted shapes and designs), stimulate a discussion, and encourage critical thinking about the efficiency and cost-effectiveness of packaging. The teacher can encourage students to answer questions such as what makes packaging efficient, does efficient packaging differ for different types of food, and what does a company have to consider when developing a cost-effective packaging for a food product? The engagement should take 5–15 minutes.

EXPLORATION

Day 2

Students will brainstorm ideas and methods for designing packaging for their products and then sketch sample designs. Students will determine the dimensions of their item to be packaged, and they must decide which three-dimensional shape will be used for the packaging. Students will keep a design notebook to keep track of their ideas and to use for discussing their final design. They will analyze a formula chart to determine which formulas could be used to calculate the dimensions of the packaging. Students use the Internet for packaging and advertisement ideas for those products. Then the teacher will encourage students to make connections between packaging and advertisements explicitly in their design notebook and to be included in their presentation. It is possible that some students may need some adjustments to the activity. As one possible accommodation, the teacher may need to provide certain students with various three-dimensional shapes and nets of those shapes to give them an idea of what they could be creating. To accomplish this part of the activity will take 30–45 minutes. Students should also be encouraged to keep track of formulas they could use as they think about potential packaging ideas.

Day 3

Students should decide on a name for their product, a design, and a logo. Students should find a name that is easy for customers to remember. The design of packaging should be eye-catching and appropriate for Conrad Foods' reputation. Students will design their packaging based on the net of their three-dimensional shapes. They will transfer their work from their design notebooks to grid paper for added precision. Students must specify the materials they are going to use for designing their classroom packaging and include this in their design notebooks along with the cost for the raw material. Remind students that if they cannot use 100% of the materials they must still include the waste in their costs. To use the materials provided students must devise a plan to determine all of the dimensions of the shape (length, width, height, circumference, radius, diameter, volume, and depending on the shape chosen, both lateral and total surface area) and record these in their design notebooks. They will base their costs on creating one package but they will compare costs based on manufacturing 75 packages. Students will maintain a log in their design notebook of their calculations for computing the costs. They will keep a comparison log to ensure that their choices are informed in the event that they do not choose the absolutely cheapest option. Student should justify their final choices and make a set of statements that can be used to convince Conrad Foods executives that their choice is a well-informed and prudent one. This part of the exploration will require 45–60 minutes.

Day 4

Students will work to construct their packaging. The teacher must ensure that each student has detailed directions contained in their design notebook. The students should be reminded that they must edit their design notebook for any changes they need to implement in the construction process. They should show what they changed when the construction started and be able to explain why they needed the changes. The design notebook should detail how parts are build as well as assembled into the package. These details will need to be provided so the factory can produce the package in quantity to the directions should be complete and accurate without missing details. Students should also need to cost out the quantity of binding materials like tape, glue, staples etc. and precisely where and how these products are used. For example, the teacher can price out and distribute 1 gram of glue or 2 linear inches of tape and then only provide that when students show their design specifications. This will also help students to cost out their package manufacturing. This can take more or less time depending on how much emphasis was placed on being specific in their design notebooks. This is best approached in small increments. So spending a 10–20 per day for three days may be a better investment and will provide time to review important mathematics skills necessary to complete the presentations.

Day 5

Students will prepare presentations (written and oral) to compare their packaging with other students' packaging. Students should make important connections from their design notebooks and their presentations. Students should be familiar with the rubrics that will be used to score their work. In their presentations, students should include what shape they selected for their packages, what material they used, why they chose that particular shape and material, and what factors they considered for making their packaging decisions. This part of the exploration will take 45–60 minutes. Allow students use PowerPoint® if they choose to and have sufficient access to computers and an LCD projector. However, there is no reason why students cannot work with the art teacher, graphics teacher, or CAD teacher to design a poster for their presentation. Expect that students will need some time to work on their presentation and if they have never been asked to prepare a presentation more time may need to be spent on the rubric so students understand the requirements and how to address all the expectations. It might be important to allow students some latitude for grammar and punctuation for non-native English speakers and the teacher should make important decisions about reasonable accommodations. For example, some students may not do well speaking in front of their classmates because of language issues related to either speech or non-native speaking. Therefore, some students may be allowed to use voice-over in their presentation or to script their presentation. It is also important to show students how to use index cards as their version of a teleprompter. Some students try to write everything on their cards while others write too little. It is important to write the main ideas to ensure that they cover all the requirements. The English teacher might address the presentation skills so it is important to check with him or her to see how you can work cooperatively to maximize the learning potential for students.

EXPLANATION

To start the project, students will have an open classroom discussion about product packaging. This discussion will engage students in the importance of packaging. The teacher should facilitate the discussion on the cost, efficiency, and attractiveness of packaging. After the discussion is over, students will start searching the Internet about packaging to discover what they will need in order to design and create their own packaging. Students should determine the important considerations for the design of packages including brand, visual appeal and design. Moreover, students should pay special attention to the material used. What is the cost of the material used? How long can the product stay in the package with that particular material? What kind of shape would be appropriate for the product?

After students gather data about their packaging, they start to plan their own packaging. Then, students make the package they design out of a material they choose. Finally, students prepare a presentation explaining their reasoning for their brand name, design, material, and shape.

EXTENSION

As an extension for this project, students can select a package that is commercially used, and they can critically analyze it to determine if it is the most efficient package for that particular product. Students must provide evidence by doing calculations to support their analysis. Students can also determine what packaging should be changed for a common product to make it more attractive to another purchaser demographic. For example, scaling up product packaging for the luxury market or scaling one down for the economy market.

APPENDIX A

Presentation Rubric 16 points

	Exemplary 4	*Mastery 3*	*Proficient 2*	*Introductory 1*
Use of visuals	Highly developed, memorable visuals illustrate main point. Manages visuals well.	Visuals clarify and illustrate main points. Presentation is memorable, and visuals are neat and well organized.	Visuals clarify and illustrate main points.	Visuals are not tightly linked to presentation and do not support or clarify main points.
Communication of Main Point	Communication is articulate and well thought out. Main points are very clear.	Communication is good; most of the main points are covered.	Communication is somewhat lacking, but main points are still evident.	Communication is poor, and main points are not fully covered.
Body Posture	Student stands tall and has confidence.	Student almost always stands tall and has confidence.	Student is nervous but is still able to convey the project.	Student is unsure what to say and is not proud of his/her work.
Fluency	Student does not stumble over words. The speech flows perfectly. Student does not use filler words.	Student does not stumble over words but uses a few filler words.	Student has a hard time getting the words out but eventually is able to communicate effectively.	Student consistently stumbles over words and uses filler words several times.

APPENDIX B

Package Rubric 20 points

	Exemplary 4	*Mastery 3*	*Proficient 2*	*Introductory 1*
Packaging	The package is most effectively designed for protecting and accommodating the food.	The package is most effectively designed for protecting and accommodating the food.	The package is somewhat effectively designed for protecting and accommodating the food.	The package was mostly empty and did not fit or protect the food.
Mathematical Knowledge	All math was solved with correct formulas demonstrating the calculations for determining the dimension for the package.	Most of the math was solved with correct formulas demonstrating the calculations for determining the dimension of the package.	An attempt was made to solve with correct formulas to demonstrating the calculations for determining the dimension of the package.	All math was solved with incorrect formulas demonstrating a lack of calculations for determining the dimension for the package.
Graphics Relevance	All graphics are related to the topic and make it easier to understand.	All graphics are related to the topic, and most make it easier to understand.	Not all graphics related, but they were easy to understand.	Graphics were hard to understand and were not related to the topic.
Attractiveness	The package is exceptionally attractive in terms of design, layout, and neatness.	The package is attractive in terms of design, layout, and neatness.	The package is acceptably attractive though it may be a bit messy.	The package is distractingly messy or very poorly designed. It is not attractive.
Participation	Student was actively engaged during the project.	Student was engaged for most of the project.	Student had to be told a few times to get back on-task.	Student was off-task for most of the project.

Crystal J. Woods
Mathematics Teacher,
Emmett J. Conrad High School, Dallas ISD

Robert M. Capraro
Teaching, Learning, and Culture
Texas A&M University

GEORGETTE JORDAN AND JAMAAL YOUNG

10. DC ELECTRONICS

Understanding series circuits and how they relate to daily use

WELL-DEFINED OUTCOME

Students will understand the societal value of electronics, be able to build a Direct Current (DC) series circuit, and calculate current, resistance, voltage, power, and speed for a car of their own design.

INTRODUCTION/BACKGROUND

Direct Current Electronics play a part in almost everything people use today. Electronics, as a field, is very broad, and it impacts every aspect of life. Every household appliance contains electronic circuits, including televisions, stereos, compact disc players, videocassette recorders, and DVD players. A meal cannot be cooked anymore without using electronics. Microwaves, toaster ovens, and even conventional ovens contain electronic sensors and controls. Beyond the essential electronic appliances are yet simpler devices, for example, laser pointers and iPods. Some electronics are powered by electricity in the form of an alternating current (AC) and others by DC. In this project, students will first explore DC circuits in series and how they can power a simple vehicle. Students will research how to build a series circuit using a lever switch (knife switch), a nine volt battery, wires, and a simple DC motor. Students will construct a car from reclaimed materials that operates by their series circuit. Upon completion of their car, they will conduct a formal presentation of their car project to the class and explain their construction methods.

OBJECTIVES

Mathematics

- The student demonstrates geometric thinking and spatial reasoning.
- The student demonstrates an understanding of two- and three-dimensional representations of geometric relationships and shapes.

Physical Science

- The student designs and analyzes series electric circuits in terms of current, resistance, voltage, and power.

Social Science

- The student compares ways humans depend on electronics and adapt to and modify the physical environment using electrically powered vehicles.

Language Arts

- The student organizes ideas for series circuit and car design in writing to ensure logical progression and support for ideas.
- The student compiles written ideas and design sketches into summaries or other formats and draws conclusions.

R. M. Capraro, M. M. Capraro, J. Morgan and J. Scheurich, (eds.), A Companion to Interdisciplinary STEM Project-Based Learning: For Teachers by Teachers, 87–94.

MATERIALS NEEDED

Needed materials may include the following:
- Lever switch
- Tape measure
- Nine volt motors with positive and negative leads
- Stopwatch
- Nine volt battery and nine volt battery snap/strap
- Hot glue and hot glue gun
- Red and black wire
- Soldering iron
- Solder
- Wire strippers/cutters
- Pulleys
- Axel material (dowel rods)
- Wheels (plastic wheels, bottle caps, blank CDs)
- Rubber bands
- Material for a car body (card board, closed cell foam).

STUDENT INTRODUCTION

Julio is interested in car electronics and wants to enhance his understanding of all the audio components he installs. He usually installs car stereos and other car electronics flawlessly, but sometimes he is forced to troubleshoot. Julio's current knowledge of electricity is rather limited; however, a better understanding of electricity may help Julio understand his installations. How can he improve his understanding of electricity and circuits? After much thought and many conversations with others, Julio decides to build a toy car that runs off of battery power. Your job is to learn about DC circuits, then design your own electronic car and prepare a presentation for the class.

ENGAGEMENT (APPROXIMATELY 1 DAY)

The teacher starts off the project by having an open discussion about series circuits and electronics in vehicles. The teacher will need to foster conversation to develop the students' interest in building their own car. Initially, students should put their ideas on paper. Tell students to take out their design notebook, and have them brainstorm about car designs (on paper, not just verbally). They may work in small groups or individually. Each teacher will have to make decisions about the availability of materials and students' interpersonal skills. Ask the students to write down anything they think they need in order to build a car. Have students turn in their design notebooks after each experience. These activities will take about half a class period. The Design Specifications should be provided early to students so they can ask questions about it and so they can start thinking in terms of parameters.

The design specifications for the project include:
- Maintain a design notebook that contains your research, car sketches, and circuit sketches.
- Include in your design notebook the calculated values for your vehicle's circuit for current, resistance, voltage, power, and speed (in both KPH and MPH; from actual trials).
- The car must be able to travel a distance of 20 meters/feet without human assistance or guidance when switched on, and if all cars meet this minimum requirement, then the fastest car will be deemed the best design.
- The vehicle's shape should be a stable three-dimensional figure and decorations are your choice.
- Vehicle weight is limited to 300 grams plus the weight of your switch and battery.
- Materials must be reclaimed (not purchased). For example, you can use cardboard, scrap wood, Styrofoam, or paper, to name a few.
- You must use the switches, wire, and batteries provided in class.
- Your total wire length cannot exceed ten inches.
- You may use the electrical tape, masking tape, duct tape, glue, markers, axels, and wheels provided in class.

– You must develop a presentation that demonstrates your design, development, and redevelopment phases; your research on creating a DC series circuit; and a comparison of the ways humans depend on electronics for modifying their physical environment.

Either continuing on or on another day use the whiteboard to list some of the things students noted about car designs. Ask students to think about the list, and have them brainstorm again so they develop more details about car design. In addition, ask students to keep a list of materials as the brainstorming session progresses. See the list of materials above for a sample list. The items in this list are suggested materials, but teachers can substitute or add materials as necessary. This brainstorming will take another half of a class period.

EXPLORATION (APPROXIMATELY 3 DAYS)

The next day re-present the ideas from the Student Introduction section above. Next, have students explore "electric cars" and "DC series circuits" on the Internet. They should make appropriate entries in their design notebook. Students will have to write down items they need in order to build their car. The final car should be constructed using only the materials listed in the design notebook. At the teacher's discretion, the students can specify additional, personal design specifications and/or performance requirements. The students will be held accountable for their own as well as for the ones specified by the teacher. If the teacher decides to let the students develop personal design specifications, each student can create a series of design specifications to constrain their car design. Some examples include: a car that only travels in a circle, monster trucks, or cars that can zigzag for a specific distance. Students will also need to carefully write down how they are going to assemble their vehicles. Figure 1 presents a sample car design (schematic). It is important for students to plan all the components. This preliminary design contains all the necessary information but is still insufficient. Students should work on their design specifications until it meets all necessary and sufficient conditions. These can be explicit in the rubrics or part of the directions the teacher gives them. These activities will take two class periods; the exploration should take three class periods altogether.

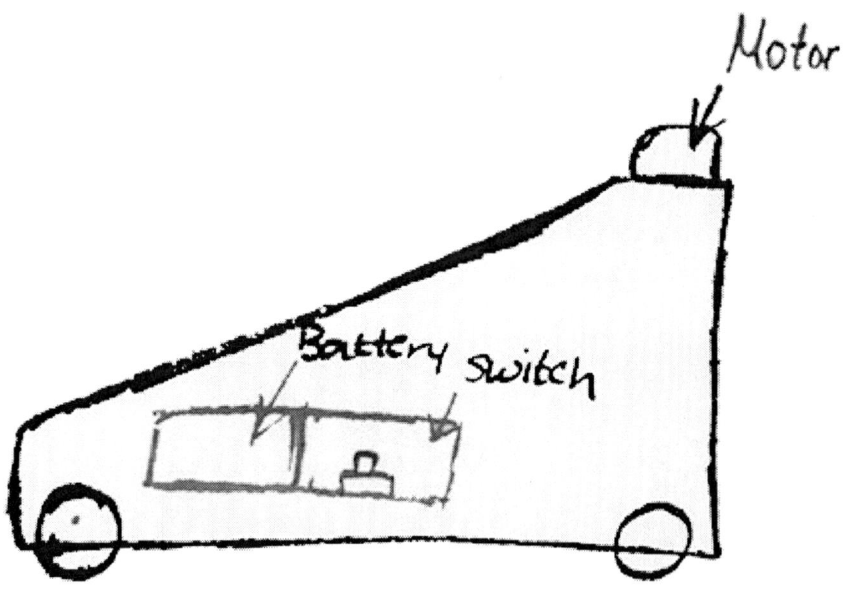

Figure 1. Design sketch.

It is important for students to make connections to the real world. The next part of the project is for students to learn how circuits impact society. Have students use the Internet, the library, or other resources to explore how electronics are used to modify the physical environment. Students also should note how important certain electronic innovations were to mastering hospitable environments in places such as Antarctica, the desert, or underwater. Students should take detailed notes in their design notebook, even if they are allowed to print from the Internet. This activity will take one class period.

EXPLANATION (3–4 DAYS)

After the students understand their task and have done initial research on the Internet about electric cars, they should sketch a preliminary car design on paper. If access to a computer laboratory with CAD software is available, this activity provides a great opportunity to have the students utilize this technology. The design is to be done individually or in pairs. However, to maintain the appropriate level of rigor, no more than two students should work on the same project. This activity will take one class period.

First, students need to draw the type of car they are going to create and be able to explain what is needed for a battery-powered car. This drawing should include a schematic diagram or a written explanation of the battery-switch operation. After students have designed their vehicle, they will work on developing a list of parts needed to assemble their cars. This list will help students understand how much of each material is needed to assemble their projects in addition to the series circuit. They will also need to meet the design parameters provided earlier (see Engagement section). Then they begin to assemble the materials needed to move their design from design to prototype. This part of the project will take one to two class periods.

Once the students begin the prototyping phase, it is important for them to design the switch first and try it to make sure it works. The functionality of the motor and its wiring may influence the design and placement of component parts in the fully developed model. The functioning model from the schematic in Figure 1 (above) is contained in Figure 2 (below). See Figure 3 for another example of a successful car.

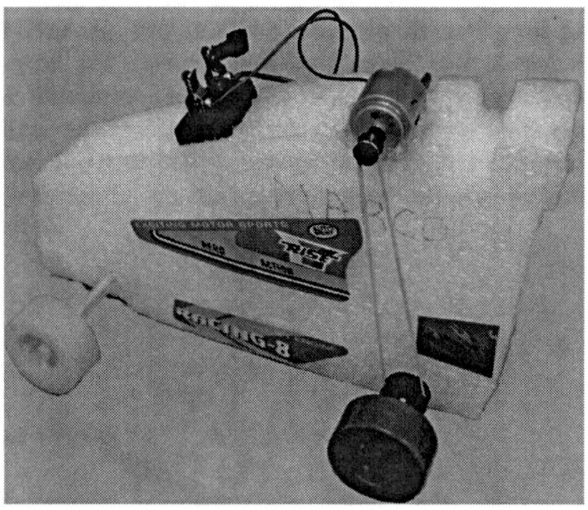

Figure 2. Car built from design schematic.

Figure 3. Sample car built from another schematic.

Once the students build the car, they will test it to make sure it meets the design specifications. They will use a distance of 20 feet as the basis for calculations. The students will collect time trial data and then calculate the car's speed based on their time and distance data.

EVALUATION

Students are evaluated based on two developmental assessments and one summative assessment. The first formative assessment addresses the student's ability to apply geometric and spatial reasoning to design and construct a physically sound design that is drawn and built to scale. Second, the teacher will assess the student's ability to calculate the car's speed and performance specifications. The final assessment is a summative assessment of the overall project; the grading rubric for the summative assessment can be found in the appendix. This rubric has a five-point scale to allow discretion. The Justification column helps students know what the teacher is thinking when he/she assigns points. This space can be used to note especially well done aspects or those aspects in which students could have done much better. It is a good idea to note specific things that would have improved the student's score. Using multiple rubrics with some overlap allows students to get a second chance to show they learned the material. For example, weighting the first two rubric scores less than the final summative one, which assesses some of the same components, allows teachers to see how students improved. When students believe that their teacher notices, they work harder, learn more, and behave better in the classroom. Giving more weight to the final rubric score helps improve not only the score but students' overall attitude as well.

EXTENSION

To extend the learning the teacher can change the gauge of the wire and have the students determine if it changes any of the important factors. For example, does the car go as fast, does it effect acceleration, battery life, or distance? What effect does a different gauge wire have on overall car weight? Does that change have any impact on overall performance of the vehicle? In addition, in mathematics classes, students can explore the various equations (e.g., d=rt, power, speed, voltage, resistance, impedance) being used and algebraically manipulate the equations to solve for different variables to answer various questions. Students can also be asked to derive various forms of equations in their physics for example, including calculations for going up or down a grade, wind resistance, and friction.

APPENDIX A

Developmental Assessment 1: Design and Construction

Category	*Criteria, 1–5 points*	*Points*	*Justification*
Design	The design contained in the notebook is buildable by a third party		
	The design reflects the schematics for powering the car		
	The design is stable and contains the necessary components for building a car		
	The materials list is complete for ALL aspects of the construction		
Construction	The actual car matches the design specifications for its construction		
	The engine (DC Circuit) is built to design specifications and located as indicated in the design specifications		
	The car is stable and rolls		
	All component parts are securely fastened in place		
	Any design modifications made on the car during production were added to the design specifications and noted		
	Total Score		

APPENDIX B

Developmental Assessment 2: Trials and Design Specifications

Category	*Criteria, 1–5 points*	*Points*	*Justification*
Trials	When switched on, the car runs, and when it is switched off, the car stops		
	The car travels at least 20 feet in a single direction three times		
	Students can answer two basic questions about the circuit they built		
	Students can troubleshoot and make modifications to the car and their design		
	Students are able to measure time and distance accurately		
Design Specifications	Students kept an accurate log of all design specifications and checked them off		
	All calculations required in the design specifications were done correctly		
	Any modifications to the car were added to the design specifications		
	Students meet personal design specifications (not required)		
	Total Score		

APPENDIX C

Summative Assessment Presentation Rubric

Category	*1*	*2*	*3*	*4*	*Total*
Organization	Audience can't understand presentation because information lacks sequence	Audience has difficulty following presentation because it jumps around	Student presents information in a logical sequence that the audience can follow	Information presented in logical, interesting sequence	
Subject Knowledge	Student does not own the information; he/she can't answer questions	Student is uncomfortable with the subject. He/she can answer simple questions but can't elaborate	Student is at ease and answers most questions with some elaboration	Student demonstrates full knowledge and answers all class questions with explanations and elaboration	
Visual Aids	Student uses superfluous visual aids or no visual aids	Student occasionally uses visual aids that rarely support the presentation	Student's visual aids relate to the presentation	Student's visual aids explain and reinforce the presentation	
Mechanics	Four or more spelling and/or grammar errors	Three misspellings and/or grammar errors	Two or less misspellings and/or grammar errors	Presentation has no misspellings or grammar errors	
Eye Contact	Student makes no eye contact and reads from notes	Student occasionally makes eye contact but mostly reads from notes	Student maintains eye contact mostly but frequently reads from notes	Student maintains eye contact with the audience and seldom reads notes	
Verbal Techniques	Student mumbles and incorrectly pronounces terms; audience members in the back can't hear	Student's voice is low, student incorrectly pronounces terms, and some audience members have difficulty hearing	Student's voice is clear. Student pronounces most terms correctly; most audience members can hear the presentation	Student's voice is clear with the correct pronunciation of terms and is easily heard by all audience members	
Group Work	Difficulty with others; students don't share decisions/ responsibilities	Works with others but has difficulty sharing decisions and responsibility	Works with the team and shares decisions and responsibility	Works very well; assumes a role in decisions and follows through	
				Total Points:	

APPENDIX D

Sample Formative Design Rubric

Task	*Poor*	*Good*	*Excellent*
Design specifications	Design does not meet the specifications	Design meets most of the specifications	Design meets all specifications
Appropriate use of scale	Inappropriate use of scale	Scale used was appropriate; minor calculation errors	Appropriate use of scale; no calculation errors
Justification	Inadequate justification	Adequate justification	Justification more than adequate; all design decisions justified
Presentation	Severely lacks detail and neatness	Adequate detail and neatness	Extreme attention to detail during presentation

APPENDIX E

Sample Formative Electric Car Scoring Rubric

	Level One	*Level Two*	*Level Three*	*Level Four*
Design	Design is partially accurate or is missing three or more appropriate components.	Design is mostly accurate and contains all but two appropriate components.	Design is mostly accurate and contains all but one appropriate component.	Design is accurate and contains all appropriate components.
Construction	Final product severely lacks components in the parts list.	Final product includes some components in the parts list.	Final product includes most components in the parts list.	Final product includes all components in the parts list.
Wiring	More than two parts of the car are not wired according to the schematic diagram.	Two parts of the car are not wired according to the schematic diagram.	One part of the car is not wired according to the schematic diagram.	The car is wired according to the schematic diagram.
Operation	The car moves erratically or in spurts.	The car neither travels in a straight line, nor does it travel the entire distance.	The car travels in a straight line but does not complete the course.	The car travels the entire required distance in a straight line.

Georgette Jordon
Electronics Teacher,
Emmett J. Conrad High School, Dallas ISD

Jamaal Young
Graduate Research Associate, Teaching, Learning, and Culture,
Texas A&M University

CHARLES MALMROS AND ROBERT M. CAPRARO

11. SETTLING THE WEST FASTER AND SAFER

AP United States History: Innovations Project

WELL DEFINED OUTCOME

Students will learn about innovations to provide human affordances during the westward march of circa 1840. They will also critically examine the production of greenhouse gases in an attempt to reduce the carbon footprint in the creation of their own innovation to help conquer the west.

INTRODUCTION/BACKGROUND

Imagine being a citizen of the United States and living in the East circa 1840. There you have most of the modern comforts: stores, mail delivery, a stage couch route, banks, telegraphs, outhouses, fresh water, a sheriff, civil government, and law and order. But just to the west lies land unsettled and dangerous with unknown perils. During this era little thought was given to carbon footprint or environmental issues, in fact, the west was considered an inexhaustible resource. For the purposes of this project "safe" is defined in the traditional way as protection of life but also more globally as managing the health of the planet and controlling green house gases.

The teacher introduces the project by saying:

> Rumor has it that there are wild men to the west known as Indians that attack settlers and steal their belongings. Even when settlers survive to establish communities these "wild men" attack, kill the settlers, and burn the town. Animals that can kill with the snap of their jaws or a toss of their tails are another danger out west. There is even a small creature know as a Widow that can kill a grown man with a tiny knick of a bite. Not to mention the bears, the mountain lions, snakes, and creatures of the waters. As President of The United States I am issuing a call to the nation to design an innovation to settle the West faster and safer than it is currently being done.

The purpose of this project is to create a representation of an innovation that helped early settlers conquer the West. As part of this project to build interdependent organized communities the students should be grouped into settlement parties (teams) of three or four. Then students should determine roles necessary for their team to develop an innovation and establish a settlement in the West. If the groups do not have three or more people, some students will have to assume more then one responsibility. The students self-determine the roles and negotiate who will assume what responsibility. This Science, Technology, Engineering, and Mathematics (STEM) PBL provides the potential for many roles; teachers may customize the roles they wish to use by defining new roles or selecting from the following sample roles:

- *Mayor (Project Manager)* Over sees everything and keeps materials in a safe place for when other team members are absent. This person is also responsible for the developing common rules and expectations.
- *Sheriff (Planner)* Helps keep consistency in the project and makes sure that each section of the innovation fits with the overall theme of the project. Enforces the rules and ensures equity for the community members.
- *Town Counsel (Designer)* Makes the project design plans for the innovation and fills out a correct form to apply for a patent.
- *Construction Engineer (Materials Manager)* Provides oversight of the materials, ensures materials are used effectively and efficiently minimizing waste/and cost.

After the roles are determined and accepted, each settler receives rudimentary provisions: (1) a piece of particleboard and each settlement party receives three pieces of parchment paper. Each settler is responsible for his or her own paper. Each student will be responsible for their own portion of the innovation, but must communicate with the settlement party as to what to add and what not to add to their portion of the design.

R. M. Capraro, M. M. Capraro, J. Morgan and J. Scheurich, (eds.), A Companion to Interdisciplinary STEM Project-Based Learning: For Teachers by Teachers, 95–103.

Students can be required to invent an item or innovation that all people (not just their settlement party) can use to help settle a rugged area quickly and safely. Students can be given the option to use an item that was invented for the settlement of the West (Circa 1840) or to create a new item that they think would have been more effective than what was actually invented. However, if students choose to create a new item, they must convince the President (teacher) that it is needed to settle the area as well as technologically plausible and feasible. It is important to have them discuss this beforehand because some students want to develop diesel or gasoline powered machinery. When this happens students can be asked to research when these were first development. For example, the first steam engine or machine was developed circa 1679 (http://wiki.answers.com/Q/When_was_the_steam_engine_invented). However, the first gasoline-powered machine was not invented until circa 1876 by Nicolaus Otto, who held several patents, (http://wiki.answers.com/Q/When_was_the_steam_engine_invented) because many of the necessary component parts had not yet been invented previously. The students must keep a design notebook; make a blueprint of the innovation and a prototype for the class to inspect. For a sample, actual pictures for the steam engine patent submission (see Figures 1 and 2) can be used. The teacher may require students to develop a design detailed enough that someone else could use it to build their innovation. This is important, and students should go the U.S. Patent and Trademark Office on the web for examples and information at http://www.uspto.gov/.

Students are required to develop a design that can be used to make their innovation, and create a prototype for the class to inspect.

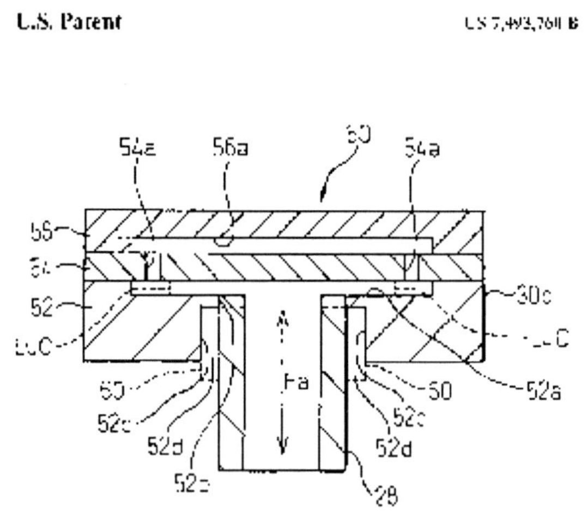

Figure 1. U.S. patent sample for a steam engine from 1876.

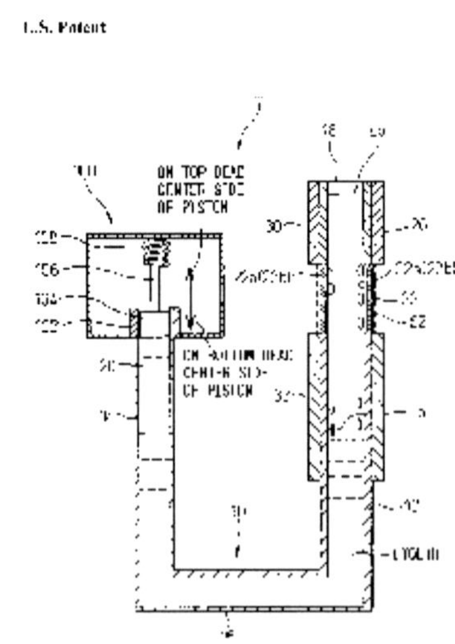

Figure 2. Sample U.S. schematics diagram for a steam engine from 1876.

Also, this project provides an opportunity for students to use CAD (computer-aided design) in their engineering class. After construction of the 3-D rendered innovation, the students' work collectively to create a presentation to try to sell their innovation to the President (teacher) and other settlers (classmates). Students should be reminded that their success in settling the west and their grade depends on people buying their innovation, and so persuasive writing and speaking skills are essential.

OBJECTIVES

Mathematics

- Spatial reasoning plays a critical role in the geometry; geometric figures provide powerful ways to represent mathematical situations and to express generalizations about space and spatial relationships. Students use geometric thinking to understand mathematical concepts and the relationships.
- Geometry consists of the study of geometric figures of zero, one, two, and three dimensions and the relationships among them. Students study properties and relationships having to do with size, shape, location, direction, and orientation of these figures.
- The student analyses geometric relationships in order to make and verify conjectures.
- The student is expected to use constructions to explore attributes of geometric figures and to make conjectures about geometric relationships.
- The student uses tools to determine measurements of geometric figures and extends measurement concepts to find perimeter, area, and volume in problem situations. Find areas of regular polygons, circles, and composite figures.
- The student uses numeric and geometric patterns to make generalizations about geometric properties, including prosperities of polygons, ratios in similar figures and solids, and angle relationships in polygons and circles.

Social Science

- The student understands the influence of scientific discoveries and technological innovations on daily life in the United States.
- The student uses problem-solving and decision-making skills, working independently and with others, in a variety of settings. The student is expected to: use a problem-solving process to identify a problem, gather information, list and consider options, consider advantages and disadvantages, choose and implement a solution, and evaluate the effectiveness of the solution; and use a decision-making process to identify a situation that requires a decision, gather information, identify options, predict consequences, and take action to implement a decision.
- The student is expected to: identify source, use, quality, and conservation of water; document the use and conservation of both renewable and non-renewable resources.

Language Arts

- The student uses recursive writing processes when appropriate.
- The student is expected to: use prewriting strategies to generate ideas, develop voice, and plan; develop drafts both alone and collaboratively by organizing and reorganizing content and by refining style to suit occasion, audience, and purpose; proofread writing for appropriateness of organization, content, style, and conventions; refine selected pieces frequently to publish for general and specific audiences; and use technology for aspects of creating, revising, editing, and publishing texts.

Science

- The student makes responsible choices in selecting everyday products and services using scientific information; evaluate the impact of research on scientific thought, society, and the environment.
- The student is expected to: make wise choices in the use and conservation of resources and the disposal or recycling of materials.

CAD

- The student knows and understands the use of the computer in helping make an innovation's design.

MATERIALS NEEDED

Scissors	Tin Cutters (for cans, it works better then scissors)
Glue Sticks (approximately 12 per class	Hot Glue Guns
Duct Tape	Regular Glue
Computers	Recyclable Materials:
Microsoft Word®	– Plastics (#s 1–5, and 7)
Rulers	– Cardboard
Three Pieces of Parchment Paper per group	– Paper
Exacto Knives	– Glass
Web Access	– Food Containers
Microsoft PowerPoint	I usually bring in some cardboard, cereal boxes and cans, for those students who just keep "forgetting" to bring in stuff, so they can at least get started.
Printer	
String	

STUDENT INTRODUCTION

The settlement parties receive a letter written by the President (See Introduction/Background) asking them to come up with ideas for innovations to help settle the west and remove any danger that settlers might face. This letter can be given to the students as a handout; groups must then come up with an idea and get it approved by the President (teacher).

Students have three days to come up with an idea and bring it back to class. Then they collect data on what the innovation will do and what it would look like.

The innovation must:

- Be composed of parts and ideas available during the time period (1840–1890's).
- Move people or things or change the physical environment.
- Be an improvement on existing technology of the time (1840–1890's).
- Be no larger than a shoebox.
- Be made in three-dimensions and comprised of reclaimed materials.
- Incorporate the real life person who invented your innovation and incorporate them into your group as a member.
- Write the history of your innovation in a paper to submit.
- Develop a presentation about your innovation, the original inventor (if applicable), how the innovation works, its design specifications, how you plan to reduce its carbon footprint, and how it made settling the west either safer or faster.

Engagement

The teacher starts off the project by showing a short video about some of the innovations that helped settle the Wild West circa 1880 using videos and DVDs from the school or the teacher's own collection. After watching the video, the class will discuss what other type of innovations could have helped the western settlers. Important questions to pose to students include: "Could we use some of these innovations today? And "How could or have the innovations changed today?" After watching the videos and engaging the students in discussion, the teacher distributes the project handout while acting out the part of the President (The letter can be constructed in part from the information provided in the Introduction/Background section). Then students use computers to do research to help them find or pick an innovation they wish to design and build (see Appendix A).

Exploration (2 Class periods)

There are two parts of exploration in this project. Students will learn about conserving the environment and wisely using their resources. First, carbon footprint is "the total set of Green House Gas (GHG) emissions caused directly and indirectly by an individual, organization, event, or product" (U.K. Carbon Trust, 2008; http://www.carbonfootprint.com/) (see Appendix C). Once the size of a carbon footprint is known, a strategy can be devised to reduce it. In this project the students will decide how much damage their innovation will do to the earth and then try to identify strategies for reducing the impact. This will get them engaged and allow them to explore the topic. Second, have the students start looking through their materials and reading the constraints to try and decide what they are going to create and what the rules will be for a suitable project. To help the students choose an innovation, start by brainstorming. Allow them to talk in their groups, choosing different ideas and then discussing them amongst the group. After they come up with a consensus on their innovation, they write it on a note card (provided by the teacher) and turn it in to the teacher. Then they can do preliminary drawings of what one of their objects might look like in their Design Notebook (See Figure 3). Measurement and patience are keys to success in this part of the project. If they want items to look realistic, they may need to find a picture and decide how they will create a scale model of the project. Students may spend a couple class periods on brainstorming about their design. It is a good time to talk about brainstorming and why it is effective. During the brainstorming, students should make notes in their design notebook and share ideas without judging ideas simply record all possibilities. Then the teacher can model a strategy for reducing the list to a single idea. In general, students should go through brainstorming on earlier class activities, perhaps dealing with, reducing the carbon footprint, so they learn that while all the suggestions will probably work to reduce their carbon footprint, not all the ideas are practical or feasible. Alternatively, the teacher can use the topics of safety and a system for distributing supplies as a brainstorm because this is important to project success and can be accomplished in a relatively short period of time to lay the foundation for how students will brainstorm.

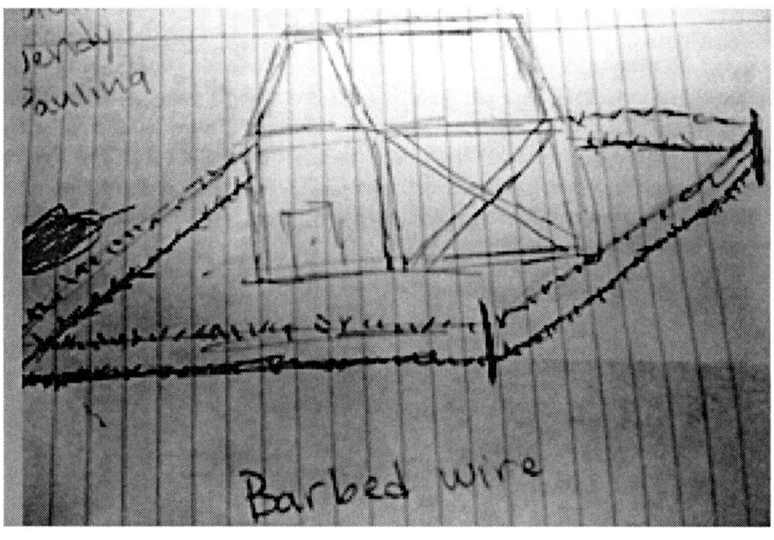

Figure 3. Drawing from the design notebook.

Explanation (1–2 Class Periods)

To start the project, students will have an open classroom discussion about biodegradable items, recycling, and the phrase "carbon footprint." (Students should be collecting recyclable items several weeks before the start of the project). After the discussions, students will log on to computers and go to the website http://www.footprintnetwork.org or type in a search engine "carbon footprint quiz." After they have the Footprint Network homepage up, have them click on "footprint for you." From there, the students will proceed to "take the quiz," which asks several questions about what size house they live in and how they get around from place to place. The teacher may print out the students' results and hang them around the room so they can compare carbon footprints. The goal is to have a low footprint, but more often than not, students want to brag about abundance. So this lesson provides an opportunity to explain why recycling and reusing is important. After the carbon footprint exercise, ask students to take out the student handout letter from the President.

Inform the students that they will be creating an innovation from the items they have collected. However, there will be constraints. Teachers should provide the constraints—either their own or the ones provided in this chapter. Teachers who would like to see more geometry included can require students to create certain objects of certain height, volume, etc.

First, the students need to consider what they have for material. They then can be divided into groups and clued in to vocabulary such as settler, settlements, and settler parties. Next, they will start to write out directions and draw preliminary sketches of what they might create. Once the brainstorming is complete (which may take one to two class periods), students will spend several class periods creating the object. The teacher's introduction to cutting and creating may take several class periods. The teacher might take time here to cover other material or introduce vocabulary because this project is designed to occur over a three-week period, bridging several important learning goals. However, I do not work on the project non-stop during that time. In fact, throughout the three weeks, I address several other topics that are important to success on the project but not sufficient for completion. After the students finish their innovations, they are to write a report about what their innovation is and how it fits into the constraints. Finally, they will present their innovation to the class. The report may be done in Microsoft Word® and the presentation in Microsoft PowerPoint®. Students tend to write more when they are required to use Word® instead of making a glitzy PowerPoint® presentation.

This part of the project will take one to two class periods.

Evaluation

The rubrics are used to help students guide their learning. Rubrics are used as both formative and summative. Students have the experience of both being rated by someone using the rubrics as well as them rating each other. When they spend the time to rate each other it improves their work. They should reflect on each rubric before it is used to provide formal feedback and grading. Have a time for students to be able to show off their innovation where it can be observed as part of their presentation. There should be a strong relationship between the sketch in their design notebook and their model (see Figure 4). The Project Design Rubric (Appendix A) should be used to evaluate the Design process. The project presentation (Appendix B) rubric can be used as summative assessment. The Project Essay rubric (Appendix D) was intended as the summative assessment to cover the well-defined outcome.

Figure 4. Model of cows, goats, pigs, and sheep contained by barbed wire.

Extension

For the extension, more advanced students can provide detailed conversions between their prototype and an original innovation. Ask them to take scale seriously (using similarity and proportions) and discuss design problems that require them to reduce large items to a small scale. The teacher can choose to account for the extension by having a supplementary set of constraints imposed on some students or groups. For example, have a student hypothesize if an object were to be used in a nanotechnology environment, what would it look like and what would it be used for? Ask students to design an innovation for exploring an environment with a gravity of one-sixth that of the earth, or ask what a tractor would look like if it was fully automated and powered by the sun.

ADDITIONAL RESOURCES INTERNET SITES

Wild West Technology http://www.tv.com/wild-west-tech/show/26187/summary.Html

HISTORY CHANNEL

WWW.HISTORY.COM/MINISITES/WILDWESTTECH

APPENDIX A

Project Design Rubric (20 points)

Category	*Exemplary 5*	*Mastery 4*	*Proficient 3*	*Introductory 2–1*
Required Elements	All materials were recyclable or biodegradable.	Mostly all materials were recyclable or biodegradable.	Only a few items were recyclable.	None of the items were recyclable/no project turned in.
Graphics & Relevance	The design fits into the required parameters of the project.	The design mostly fits the requirements of the project.	The design follows only a few of the requirements.	The student's object was incomplete, and no constraints were followed.
Attractiveness	The object was original. Students worked hard to make it realistic. Overall appearance looked clean and neat with sturdy construction.	The object was sharp and clean. Students worked hard and put in a lot of effort. Parts appeared to be attached ok. Not very realistic.	The object was somewhat sloppy but identifiable.	The object was sloppily constructed. It appeared to be or was falling apart. Effort was not evident. Not sturdy.
Participation	Student was actively engaged during the project.	Student was engaged most of the project.	Student had to be told to get back on task.	Student was off task frequently and required redirection.

APPENDIX B

Presentation Rubric (16 points)

Category	*Exemplary 4*	*Mastery 3*	*Proficient 2*	*Introductory 1*
Visuals	Highly developed, memorable visuals illustrate main point. Manages visuals.	Visuals clarify and illustrate main point and are neat and well organized. Presentation is interesting.	Lack of or poor choice of visuals or not closely linked to main points.	Visuals not linked to presentation or distracting; they do not support or clarify main points.
Communication	Communication is articulate and planned. The student does not depend on slides and makes eye contact. Main points are clear.	Communication is good; most of the main points are covered. Some dependence on the slides. The student makes eye contact frequently.	Communication is lacking, though main points are made. Too much text on slides; little eye contact.	Communication is poor, and main points are not evident. The student exhibits an over-reliance on slides and no or little eye contact.
Body posture	The student stands tall, shows confidence, and is familiar with the presentation and its organization.	The student almost always stands tall and shows confidence, but sometimes talks to the presentation or floor.	The student is nervous and talks too quietly but conveys the presentation.	The student is clearly nervous and unsure of the presentation, talks to the presentation or floor, and is hard to hear.
Presentation fluency	The student makes minimal pauses, avoids filler words (um, ah, hums), and handles disruptions or questions.	The student pauses occasionally, uses 2–3 filler words, does not stumble on words, and is sidetracked somewhat by questions.	The student pauses frequently, has difficulty with words, uses 4–5 filler words, and is easily sidetracked.	The student consistently stumbles on words, uses filler words frequently, and is not familiar with the presentation.

APPENDIX C

Recyclable Items (16 points; counts twice for 32 points)

	Exemplary 4	*Mastery 3*	*Proficient 2*	*Introductory 1*
Required Elements	All materials were recyclable or biodegradable.	Mostly all materials were recyclable or biodegradable.	Only a few items were recyclable.	None of the items were recyclable.
Graphics & Relevance	The object fit into one of the constraints, and the student could explain how and why.	The object fit into one of the constraints, but the student had a difficult time explaining how and why.	The object did not fit into one of the constraints.	The object was incomplete, and no constraints were followed.
Attractiveness	The object was original. The students worked hard to make it realistic. The overall appearance was clean and neat.	The object was sharp and clean. The students worked hard and put in a lot of effort. Parts appeared to be together ok.	The object was somewhat sloppy, but the students got the point across.	The object was sloppy. It appeared to be falling apart. Effort was not shown by its attractiveness.
Participation	The student was actively engaged during the project.	The student was engaged during most of the project.	The student had to be told a few times to get back on task.	The student was off task most of the project.

APPENDIX D

Project Essay Rubric

	Exemplary 4	*Mastery 3*	*Proficient 2*	*Introductory 1*
Language Use	Highly developed essay with a thesis and conclusion and 0–2 mistakes in writing.	Well-developed essay with a thesis and conclusion and 3–6 mistakes in writing.	Essay is developed well and includes either a thesis or conclusion and no more than 10 mistakes in writing.	Essay is not developed and does not include either a thesis or conclusion.
Communication	Communication is articulate and well thought-out. Main points are very clear.	Communication is good, and most of the main points are covered.	Communication is somewhat lacking, but the main points are still evident.	Communication is poor, and main points do not appear to be fully covered.
Focus	The student explains and shows how other innovations may have been affected.	The student stays on task in the essay while explaining his/her innovation.	The student talks about his/her project but goes off topic at times.	The student is not sure what he/she is writing about.
Format	The essay is made up of five paragraphs.	The essay is made up of four paragraphs.	The essay is made up of two to three paragraphs.	The essay is only one paragraph.

Charles "Chuck" Malmoros
History Teacher,
Emmett J. Conrad High School, Dallas ISD

Robert M. Capraro
Teaching, Learning, and Culture,
Texas A&M University

LISA R. W. COBB AND KIMBERLY E. MCMILLAN

12. THE LION LAUNCH

WELL-DEFINED OUTCOME

Students will research and construct a working siege machine to accurately predict the trajectory of a projectile through analyzing the energies and forces at work in the student-built machine.

INTRODUCTION/BACKGROUND

In order to understand the purpose and function of a given machine, detailed analysis must be conducted to identify and understand the forces and energy transformations involved in the action of the machine. Students will explore these concepts through this project-based learning (PBL) experience. Students will be expected to research medieval war machines, build a working model, explain the mechanics behind its function, and predict the trajectory of a projectile launched by the machine. Students will enter their research information and experimental design in a journal and then will build a working model with the objective of accurately predicting the trajectory of the projectile launched from the machine.

This project integrates the study of one-dimensional kinematics, two-dimensional kinematics, forces, and energy conservation and transformations. Students have experienced these topics through multiple classroom experiences, as well as in their daily lives. Through the use of a small, disposable model, students will be able to infer the proportions and relationships of the mass, force, and energy components of the machine. The students will then apply this understanding to build their own table-top model capable of launching a projectile at least one meter.

OBJECTIVES

Mathematics

- Collect and organize data, make and interpret scatter plots, fit the graph of a function to the data, interpret the results, and proceed to model, predict and make decisions and critical judgments.
- Relate representations of quadratic functions, such as algebraic, tabular, graphical, and verbal descriptions.
- Use functions and their properties, tools, and technologies to model and solve meaningful problems.
- Use vectors to model physical situations.

Physical Science

- Analyze examples of uniform and accelerated motion including linear and projectile.
- Demonstrate the effects of forces on the motion of objects.
- Observe and describe examples of kinetic and potential energy and their transformations.
- Calculate the mechanical energy in a physical system.
- Demonstrate the conservation of energy.

Language Arts

- Determine the meaning of grade-level, technical academic English words in multiple content areas.
- Analyze factual, quantitative, or technical data presented in multiple graphical sources.
- Compare and contrast how events are presented and information is communicated by visual images versus non-visual texts.

R. M. Capraro, M. M. Capraro, J. Morgan and J. Scheurich, (eds.), A Companion to Interdisciplinary STEM Project-Based Learning: For Teachers by Teachers, 105–116.

- Produce a multimedia presentation with graphics, images, and sound that conveys a distinctive point of view and appeals to a specific audience.
- Organize information gathered from multiple sources to create a variety of graphics and three dimensional forms.
- Paraphrase, summarize, quote, and accurately cite all researched information according to a standard format.
- Participate productively in teams; build on the ideas of others, contribute relevant information, develop a plan for consensus building, and set ground rules for decision making.

MATERIALS NEEDED

- Medieval war task scenario
- Internet access
- Journals
- Grading rubrics
- Possible model materials: chopsticks, toothpicks, popsicle sticks, masking tape, rubber bands, paper clips, string, straws, fishing weights, various batteries, metal washers, nuts and bolts, precut wood (size depends on design), hammers, nails, screws, wood glue, PVC pipe, metal re-bar, metal pipes

Time frame for the project. The project should last one week.

DAY 1

Engagement

King Richard the Lionhearted has called all craftsmen in England to design and build mighty war machines for use on his military exploits. He has traveled to faraway lands and has encountered many nations, all with different war technology. He needs siege machines that will give him the best military advantage in battle. King Richard's army must be able to demolish enemy fortifications. He charges you with the task of determining what kind of machine will do the most damage (by launching a projectile along a desired trajectory) and yet be adaptable to various battle scenarios.

Students will break into groups of three or four once the teacher has read the engagement. The students will establish roles and responsibilities and plan how they will proceed with this task. The roles and responsibilities must be recorded in each team member's journal (see Table 1).

Table 1. Sample journal entry

Name	*Roles & Responsibilities*
Joe	Leader/Editor
Bob	Materials Manager
Marisol	Recorder
Suzy	Spokesperson

Exploration

King Richard's High Commander explains that there are several different types of machines that can launch a projectile in battle. He charges you with the task of determining which kind of machine will cause the most damage and give King Richard the military advantage. The machine chosen must be able to be moved around the battlefield and must be able to be modified to launch the projectile varying distances.

Students will research, compare, and contrast the different kinds of siege weapons and record the information in their journal. Students may utilize a table format or concept map to show the similarities and differences between the machines. Then, students will make a hypothesis about which machine will be the most effective in performing the task for King Richard. This hypothesis will be recorded in their journal. (See the journal checklist shown in Table 2 and Appendix A.)

Table 2. Journal checklist

Journal Checklist	*Day 1*
Roles and responsibilities recorded for all members	2 1
Comparison of machines present	2 1
Concept map correctly used to compare/contrast	5 4 3 2 1
Hypothesis predicting best machine choice	5 4 3 2 1
Justification of chosen machine	5 4 3 2 1
Schematic of the machine is drawn with parts labeled	2 1
Schematic includes accurate description of forces and energy transformations	5 4 3 2 1
Schematic includes sizes of components and proportions involved	2 1
Proportion hypothesis is explained in writing	2 1
TOTAL	/ 30 POINTS

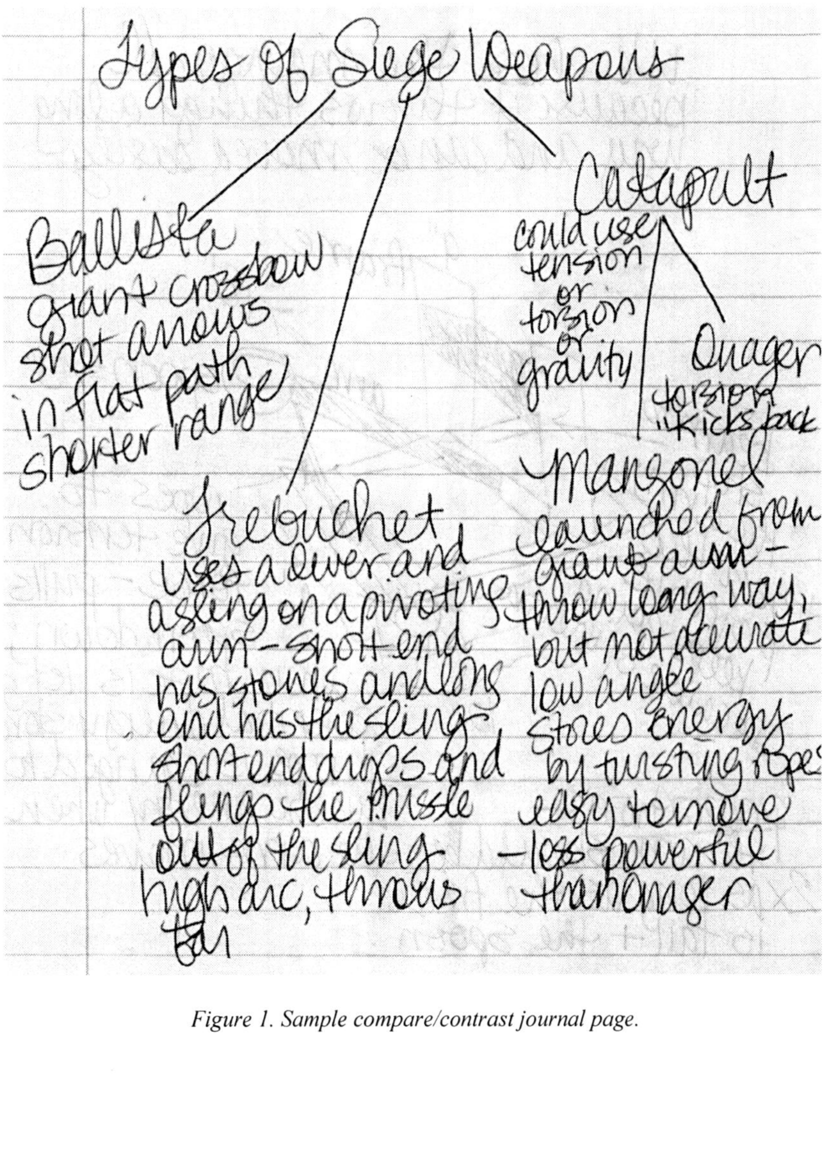

Figure 1. Sample compare/contrast journal page.

Students will research their machine of choice to discover the design parameters and the forces and energy transformations inherent in its functioning. Students will justify in writing their machine decision and will create a drawing in their journal (see Figures 1 and 2) identifying the parts of the machine, the forces involved, and the energy transformations that take place during the firing of the machine. Students will indicate on the drawing the proposed size for each component of the machine. Students will also explain in writing the proportions of the throwing arm compared to the frame and counterweight (if used) that they believe will function reliably. The written and journal components may be finished as homework if not completed in class.

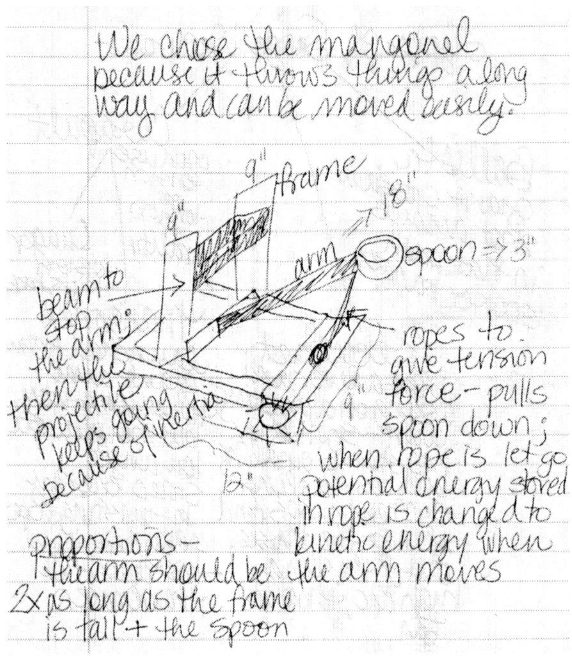

Figure 2. Sample journal entry with proposed machine information.

DAY 2

Using the provided materials, student groups will create a small model of their chosen machine that is no larger than two square feet. Students will experiment with the predicted proportions hypothesized on Day 1 and modify as necessary until they have a working model, which they will draw and label in their journal. The group must demonstrate their machine's performance by launching the given projectile at least one meter. The construction of the model will be formatively assessed using the construction rubric (see Appendix B).

DAY 3

Students will finish building their machine and perform mathematical calculations to prove the launching capacity of the machine. Students' model construction will be formatively evaluated using the provided rubric, available in the Appendix B. Students will utilize the provided lab report (see Appendix C) to record the experimental parameters. The lab report may be finished for homework if not completed in class.

DAY 4

Explanation

Students will prepare a PowerPoint presentation that will include the following components:

- Findings on the different types of siege machines
- Justification of the group hypothesis for the best war machine
- Explanation of the forces and energy transformations used in the chosen machine

- Pictures or drawings of the machine with proportions and sizes included
- Procedures for moving and modifying the machine

Students will be formatively assessed (see Appendix D) using a checklist of components in the PowerPoint presentation.

The history. Early modern warfare developed from the mid-15th century through the end of the 18th century. During the Middle Ages, chivalry and siege warfare both played an integral part of the power struggles of Europe and the Middle East. Attackers trying to acquire new lands first tried to negotiate terms of surrender before besieging the town or castle. If terms were not agreed upon, the attacker would forcibly advance and try to destroy the fortifications of the town or castle. In order to do this, the attackers used mighty siege machines to destroy the walls and kill or demoralize the inhabitants of the town. Students will gain an understanding of the technology of the time period and relate that to the physical concepts inherent in the functioning of the machines.

The physics and mathematics. Siege weapons required the use of various physical forces to cause devastation to fortified walls. Some machines used the force of gravity to transform gravitational potential energy into the kinetic energy of the projectile. The catapult used a counterweight on the short end of the throwing arm to force a projectile at the long end to fly through the air. The trebuchet used the lever action of a throwing arm with a leather sling on the long end and a counterweight on the short end, which forcibly launched a projectile out of the sling. Using the principles related to two-dimensional and projectile motion, several physical quantities will be calculated and used to analyze and predict the trajectory of a projectile. In this way, a machine's efficiency for a specific need on the battlefield can be demonstrated.

DAY 5

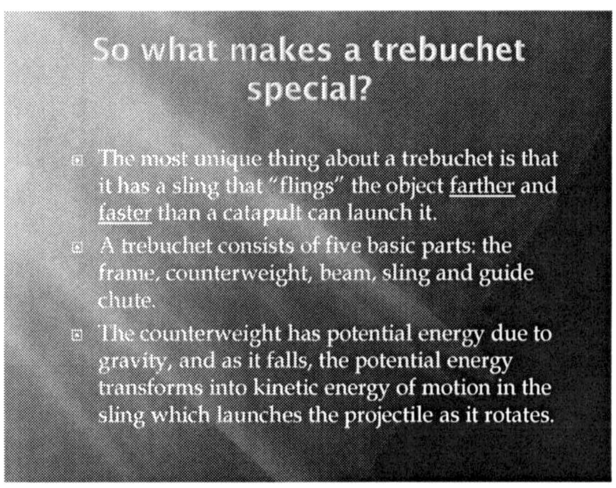

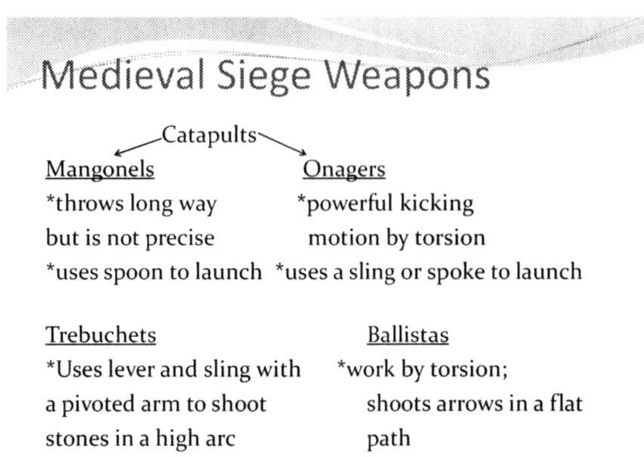

Figure 3. Example PowerPoint slides.

Evaluation

Students will present their proposal to King Richard and his High Commander. Each group will be given five to seven minutes to present and demonstrate the launching capability of their model. Students and/or teachers may wish to dress in medieval attire to enhance the presentation. The summative rubric (see Appendix E) will be used to assess each proposal. Example slides are included below in Figure 3.

Extension

Students will have a choice of extension activities: a written report or an advertisement promoting their war machine. If the written report is chosen, students will compare and contrast their chosen medieval siege machine with currently used weaponry such as cannons and tanks. The report will be a two-page, typed essay evaluated based on the rubric provided in Appendix F. If the advertisement is chosen, students will promote and market their machine through audiovisual presentations. Evaluation for the promotion of the war machine will follow the rubric found in Appendix G.

APPENDIX A

Journal Checklist

Journal Checklist	*Day 1*
Roles and responsibilities recorded for all members	2 1
Comparison of machines present	2 1
Concept map correctly used to compare/contrast	5 4 3 2 1
Hypothesis predicting best machine choice	5 4 3 2 1
Justification of chosen machine	5 4 3 2 1
Schematic of the machine is drawn with parts labeled	2 1
Schematic includes accurate description of forces and energy transformations	5 4 3 2 1
Schematic includes sizes of components and proportions involved	2 1
Proportion hypothesis is explained in writing	2 1
TOTAL	/ 30 POINTS

APPENDIX B

Construction Rubric

CATEGORY	***4***	***3***	***2***	***1***
Construction Materials (No kits used)	Appropriate materials were selected and creatively modified in ways that made them even better.	Appropriate materials were selected and there was an attempt at creative modification to make them even better.	Appropriate materials were selected.	Inappropriate materials were selected and contributed to a product that performed poorly.
Construction Care Taken	Great care was taken in the construction process so that the structure is neat, attractive, and follows student design accurately.	Construction was careful and accurate for the most part, but 1–2 details could have been refined for a more attractive product.	Construction accurately followed the plans, but 3–4 details could have been refined for a more attractive product.	Construction appears careless or haphazard. Many details needed refinement for a strong or attractive product.
Function	Structure functioned extraordinarily well, holding up under atypical stresses.	Structure functioned well, holding up under typical stresses.	Structure functioned pretty well, but deteriorated under typical stresses.	Fatal flaws in function with complete failure under typical stresses.
Data Collection and Analysis	Data was taken several times in a careful, reliable manner and analyzed with less than 2 calculation errors.	Data was taken twice in a careful, reliable manner and analyzed with 3–5 calculation errors.	Data was taken once in a careful, reliable manner, and an attempt was made to analyze and calculate requested quantities.	Data was not taken carefully OR not taken in a reliable manner. No attempt was made to analyze the data.

Construction Rubric (continued)

Scientific Knowledge	Explanations by all group members indicated a clear and accurate understanding of the scientific principles underlying the construction and modification process.	Explanations by all group members indicated a relatively accurate understanding of the scientific principles underlying the construction and modification process.	Explanations by most group members indicated a relatively accurate understanding of the scientific principles underlying the construction and modification process.	Explanations by several members of the group did not illustrate much understanding of the scientific principles underlying the construction and modification process.

APPENDIX C

Lion Launch Lab Report

Make your measurements and calculations using the following equations to calculate the requested quantities when you test your model siege weapon.

I. Time in air and horizontal displacement measured during firing
**Use the timer and tape measures to experimentally gather your data for at least 3 flights of your machine.

Trial	***Time in Air***	***Horizontal Displacement***	***Counterweight or tensional material description***	***Height of counterweight (if applicable)***
1				
2				
3				

II. Initial horizontal velocity calculations
**Use the data above to calculate the horizontal average velocity for each of your trials.

Trial	***Calculation using v = d / t***	***Horizontal Velocity***
1		
2		
3		

******<u>NOTE</u>: Now choose your best flight from the data above and use ONLY that data for the rest of the calculations!******

III. Initial vertical velocity calculation

Step 1. Experimental time / 2 = time to the peak of the projectile
Step 2. Now take the time from above, and plug it into the following equation and solve for v_{iy} ($g = 9.8$ m/s^2):

$$t_{up} = v_{iy}/g$$

v_{iy} = ________________

IV. Maximum height calculation (*y*)
**To find the maximum height, use the following equation:

$$y = v_{iy}\, t_{up} + \tfrac{1}{2}\, g\, t_{up}^{2}$$

y = ________________

V. Initial total velocity calculation (v)
**To find the total velocity, you must add the horizontal and vertical component vectors v_{iy} and v_{ix}.

<u>REMEMBER</u>: Velocity is a vector, so you must use vector addition to find the total velocity!

v_{total} = ______________________

VI. Kinetic Energy
**To find the kinetic energy of the projectile, use the equation below with your total velocity (m = mass of the projectile, v = velocity of projectile):

$$KE = \frac{1}{2} mv^2$$

KE = ______________________

VII. Potential Energy
**To find the gravitational potential energy of the counterweight, measure the height of the counterweight above the ground and plug that measurement into the GPE formula below where m = mass of the counterweight, h = height of counterweight above the ground, and $g = 9.8\ m/s^2$. If you have an elastic material providing potential energy, then you would use the equation on the right, where k is the spring constant and x is the amount of compression. You may or may not be able to find the spring constant for your material. If not, estimate to the best of your ability and show how you came to your conclusion.

$$GPE = m\ g\ h \qquad \text{or} \qquad PE_{elastic} = \frac{1}{2}\ k\ x^2$$

PE = ______________________

VIII. Initial launch angle calculation
**Use the following formula to find the launch angle:

$$\Theta = \tan^{-1} \frac{(v_{iy})}{(v_{ix})}$$

Θ = ______________________

IX. Range calculation (x)

$$x = v_{ix}\ (t)$$

x = ______________________

X. Reflection on the test flight process
Record here your thoughts about this activity including both the good and bad aspects of your flights and the calculations.

XI. Modification Plan
How could you modify your machine to make it function better? Explain how you would modify, specifically noting what outcome you would expect.

APPENDIX D

PowerPoint Presentation Checklist

PowerPoint Components	*Check off when done*
Findings on the different types of siege machines	☐
Justification of hypothesis for the best war machine	☐
Forces and energy transformations used in the chosen machine	☐
Pictures or drawings of the machine with proportions and sizes included	☐
Procedures for moving and modifying the machine	☐

APPENDIX E

Summative Rubric

CATEGORY	*4*	*3*	*2*	*1*
Scientific Accuracy	All scientific concepts are clearly and explicitly explained.	Most scientific concepts are explained with comprehension.	Some scientific concepts are discussed.	Minimal scientific explanation is given.
Content	Presentation contains all components to depth and complexity.	Presentation contains all components, but some are only superficially addressed.	Presentation is missing 1–2 components, but others are adequately explained.	Presentation is lacking in coherence and is grossly missing the required content.
Creativity	Presentation shows great creativity and ingenuity through slide design and presentation skills.	Presentation is well done and utilizes some creative elements.	Presentation is complete yet is lacking in creative flair.	Presentation displays no creativity and only portions of the content.
Machine Performance	Model performs above and beyond the requirements with no breakdown in structure.	Model performs sufficiently to meet the 1 meter launch distance.	Model performs but falls short of the distance and/or falls apart when launched.	Model does not perform when required during the presentation.
Presentation	Presentation meets the time frame and is conducted in a clear and concise manner.	Presentation meets the time frame but lacks clarity and voice projection.	Presentation is slightly outside of the time frame, but is clear and concise.	Presentation is grossly outside of the time frame and has an unclear and sloppy delivery of information.

APPENDIX F

Extension Rubric for the Written Report

CATEGORY	*4*	*3*	*2*	*1*
Quality of Explanation	Explanation of the war machine is clear, accurate, and easy to understand.	Explanation is mostly accurate and simple to understand.	Some of the explanation is accurate but is difficult to follow.	Explanation is unclear and incorrect.
Comparison	Student appropriately compares and contrasts the types of machines and justifies their choice.	Student compares or contrasts the types of machines and justifies their choice.	Student compares the types of machines.	Student inaccurately compares the types of machines.
Content Knowledge	All information in the proposal is accurate.	Most of the information in the proposal is accurate.	Some of the information in the proposal is accurate.	The proposal contains incorrect information.
Historical Background	Three historical facts are presented, defended, and properly cited.	Two historical facts are presented, defended, and properly cited.	One historical fact is presented and properly cited.	No historical information is presented, or citations are incorrect.
Format	Paper has 2 typed, double-spaced pages in 12 point Times New Roman font with 1" margins and a cover page.	Paper has 2 typed, double-spaced pages, and a cover page is attached.	Paper has 1–2 typed, double-spaced pages, but there is no cover page.	Paper is 1–2 pages without a cover page.

APPENDIX G

Extension Rubric for the Promotion of the War Machine

CATEGORY	*4*	*3*	*2*	*1*
Visual Quality	Project is clean, neat, colorful, and visually appealing.	Project has some color and is neat but not visually stimulating.	Project is complete but lacks color and inspiration.	No project or messy, incomplete project.
Quality of Explanation	Explanation of war machine is clear, accurate, and easy to understand.	Explanation is mostly accurate and simple to understand.	Some of the explanation is accurate, but it is difficult to follow.	No project or explanation is unclear and incorrect.
Persuasive Appeal	Presentation of the war machine is convincing and appealing.	Presentation sways the audience but is not fully believable.	Presentation does not win over the audience.	No project or lacks inspiration or persuasive creativity.

Extension Rubric for the Promotion of the War Machine (continued)

Content Knowledge	All information in the proposal is accurate.	Most of the information in the proposal is accurate.	Some of the information in the proposal is accurate.	No project or the proposal contains irrelevant information.
Time	Presentation is exactly 5–7 minutes.	Presentation is too short or too long by 1 minute.	Presentation is too short or too long by 2 minutes.	No project or presentation of project is 1–2 minutes long.

Lisa R. W. Cobb
Science Instructional Specialist,
Waco High School, Waco Independent School District

Kimberly E. McMillan
Secondary Science Specialist,
Waco Independent School District

THE WORLD AROUND US

WILLIAM PAUL AND MARY MARGARET CAPRARO

13. THE DEPRESSION NOW AND THEN

WELL-DEFINED OUTCOME

Students will examine information from the Great Depression to make predictions about the outcome of the current economic condition.

INTRODUCTION/BACKGROUND

The economic condition of a country is an important topic to examine because of its effects on job markets and salaries, commercial property and home prices and sales, industry and manufacturing, bank rates, and the stock market. Students will explore these and other related topics as they work through this project-based learning activity. Students will be expected to research and reflect on the causes and effects of the Great Depression and relate these historical events and perspectives to the current economic conditions in the United States. Students will enter their findings in a journal and then prepare and present a PowerPoint presentation that illustrates connections between the Great Depression and current economic conditions.

This project was designed during the 2008–2009 school year. The United States had been experiencing a deep recession, and the causes were comparable to the causes of the Great Depression. This project was designed for seniors who would have already taken United States history. These particular students should have some background on the causes of the Great Depression and how the government responded through the New Deal. The students should see that there is a parallel between the causes and should study the response of the New Deal programs to gain insight as to what they could recommend to the current government. Several of the students came up with responses very similar to that of Herbert Hoover and Franklin Delano Roosevelt.

OBJECTIVES

U.S. History

- Analyze causes of economic growth and prosperity in the 1920s, including the impact of World War I on the U.S.
- Analyze political, economic, social, and demographic data to determine the level of development and standard of living of the nation.
- Analyze information by sequencing, identifying cause-and-effect relationships, finding the main idea, summarizing, making generalizations and predictions, and drawing inferences and conclusions.
- Use primary and secondary sources such as media and news services, biographies, interviews, and artifacts to acquire information about the U.S.
- Analyze the causes of the Great Depression, including the decline in worldwide trade, the stock market crash, and bank failures.

Economics

- Evaluate economic-activity patterns using charts, tables, graphs, and maps.
- Create a product on a contemporary economic issue or topic using critical methods of inquiry.
- Use appropriate mathematical skills to interpret social studies information.

Mathematics

- Represent relationships among quantities using graphs.
- Interprets situations in terms of given graphs.

R. M. Capraro, M. M. Capraro, J. Morgan and J. Scheurich, (eds.), A Companion to Interdisciplinary STEM Project-Based Learning: For Teachers by Teachers, 119–126.

Language Arts

- Organize ideas in writing to ensure coherence, logical progression, and support for ideas.
- Compile written ideas and representations into reports, summaries, or other formats and draw conclusions.

Sciences

- Communicate and apply scientific information extracted from various sources such as current or historical events, news reports, published journal articles, and marketing materials.
- Research and describe the history of science, contributions of scientists, and impact on society.

MATERIALS NEEDED

- Economic advisor scenario
- Internet and library access
- Microsoft PowerPoint
- Grading rubrics
- Journals

Time frame for the project. The project should last one week.

DAY 1

Engagement

After the teacher reads the engagement below, students break up into partners to discuss how they will get started on the charge given to them.

For the last 15 years, Franco Modigliani has been considered the top economic analyst in the United States. He has been working for the Econ Group, a Dallas-based think tank that contracts itself out to developed and developing nations as a consultant to provide solutions to economic problems.

On the evening of November 5, one day after the 2008 presidential election, Franco received a phone call to his office. On the phone was none other than the President-elect, Barack Obama. Obama started off with small talk but eventually got to the point of the phone call:

> I need a solid economic advisor. My people have told me you are the best in the business. As you know, our current economic situation is dire, and I need a good problem solver in this position. Should you accept the position, you will be able to bring along two assistants. The position starts on Month, Day, and Year. What do you think?

Franco immediately agreed and upon hanging up the phone already had two people he knew would be great assistants to him. Find a partner and begin discussing how you will help Franco advise the President.

Exploration

Upon their arrival in Washington D.C., Franco and his team are given a tour of the Capitol and entertained. At the end of the tour, they are taken to their office near the White House and told that work begins tomorrow morning at 7:00 sharp. After a night's sleep, the team arrives at the White House conference room to meet with the new president, who gives this introductory speech:

> Ladies and gentleman, as you are all aware, we have an economic crisis on our hands. We are currently heading into one of the deepest recessions since the Great Depression. I am expected to give a presentation to Congress on the state of the current U.S. economy. Many don't seem to comprehend the total devastation we are currently experiencing in this country. What would probably get their attention is a comparison between the causes of our current economic state and the causes of the Great Depression. I would also like to see the effects of the Depression alongside the current effects we're experiencing, along with any recommendations you may have for our problems. I need this in a week.

The President leaves the team to work, and as the group heads back to the office, Franco mentions he would like to see the information prepared in a PowerPoint format (other formats that could be considered are a poster, monologue, or a play). The teacher draws a chart on the board, and students create the same chart in each of their journals. The students will engage in a teacher-led discussion to fill in the chart. An example is provided below in Table 1:

Table 1. Sample journal entry

Great Depression Causes	*2008 Recession Causes*
Speculation Too much credit Lack of regulation Borrowing on the margin 1929 Crash No bank insurance "Bank runs"	Speculation Too much credit De-regulation Bad mortgage loans

DAY 2

Students will break into groups and access the library, books, and/or Internet to research in detail the causes of the Great Depression and the current recession. They will be recorded in the table below.

Table 2. Example of a blank journal format

Great Depression Causes	*2008 Recession Causes*
* * * * *	* * * * *

The teacher will assess student journals using the informal checklist shown in Table 3.

Table 3. Journal checklist

Journal Checklist	
Setup of chart is accurate (title, columns)	5 4 3 2 1
Five unique Great Depression causes are included	5 4 3 2 1
Five unique 2008 Recession causes are included	5 4 3 2 1
TOTAL	/15

DAY 3

Explanation

The students are to prepare a PowerPoint presentation that will include five items:

- Three Great Depression/2008 Recession causes. The causes should be similar to provide a basis for understanding the current situation.
- Two effects of the Great Depression/2008 Recession. The effects should be similar to provide a basis for understanding the current situation.
- Two actions by the government to alleviate the effects in each situation.
- Two reforms instituted by the government in the post-Depression era.

- Two government/private reforms needed in order to prevent post-2008 problems.
- Comparison of the physical state of citizens before and after the Depression with emphasis on major diseases and the availability of medical care (optional).

The History

History teaches the Great Depression occurred because of a lack of responsibility by Wall Street, self-control by the general population, and government oversight. What should have been a natural decline in the business cycle turned into an economic depression. The Great Depression and subsequent eras led to the creation of a quasi-welfare state where the general public came to expect the government to impose regulations on businesses. Because of the Great Depression, the United States now has the Securities and Exchange Commission (SEC), Federal Deposit Insurance Corporation (FDIC), deficit spending, and Social Security, as well as significant government regulation. Through this project, students should gain some understanding of how these governmental agencies are tied to economic policy.

Economics

The economic aspect of the project will produce an enhanced understanding of the business cycle, free market principles, and how deregulation has led to business carelessness. (The students that are taking Economics could be assigned to pick a stock and pretend to buy 100–1000 shares of that stock. They should report out at the end of this project or in a month to 2 months time how the stock did, and try to explain why they think the stock performed the way it did). Students should learn to read stock via the newspaper instead of using the Internet. At the end of the stock market part, students can create charts and graphs based on each person's stock. So if you were to create a bar chart then on the left would be the dollar amounts and on the bottom would be the stock name. So students could compare their stock data as a whole. Or use a scatter plot for rate of change of their stocks.

Government

The student will gain an understanding of the government's historical role in shaping economic policy and how this knowledge can be used to predict the government's current economic policy.

The students will use their research skills to gather information and begin organizing their findings into a PowerPoint presentation. The Day 3 assessment is based on student cooperation, participation, and engagement in planning group PowerPoint presentations. Both the teacher and students will use the Participation and Cooperation rubric (see Appendix A) to evaluate their team member in order to ensure that both members of the group participate in gathering the information necessary for their presentation.

Medicine

During the Depression there was a shift in the type of medical care provided based on income and the ability to pay for services. Those people that were considered poor before the depression and remained poor afterwards had the largest proportion of free care compared to higher income groups. They also received the most hospital and nurse care. Financially comfortable people received more actual physician attention, largely due to the fact that they were often able to pay for the services. Physicians provided the majority of medical services themselves, and the availability of nurse or hospital supportive care was not common. Tuberculosis was prevalent along with other diseases resulting from lower standard of living of most Americans due to the economic instability of the nation.

DAY 4

The students will work on completing their PowerPoint presentations by checking spelling, grammar, and background. They will practice their presentations within their groups for determine if they have covered all the necessary content. Group members will use the final rubric to critique each other for practice.

A formative assessment will be used to check for students' progress towards readiness for their final PowerPoint presentation. Group journals should include a chart that includes the following information (see Figure 1 and Table 4). A student example of a journal page is displayed below (Figure 2). A checklist to assess journal entries is included in Appendix B.

- 3 causes of the Great Depression
- 3 causes of the current recession
- 2 effects of the Great Depression
- 2 effects of the current recession
- 2 government actions taken to alleviate the Great Depression
- 2 government actions taken to alleviate the current recession
- 2 reforms implemented in the post-Depression era
- 2 reforms YOU recommend the current government to make to prevent future recessions
- Comparison of health and the level of medical care accessibility (optional)
- Include: relevant pictures, charts, graphs, etc.

All information used in your project is borrowed information. Therefore, it is essential that you document within the PowerPoint who the information was created by. You MUST give credit to the original author. Not only is this the legal way to make a presentation, it also gives your ideas credibility.

Figure 1. Presentation content.

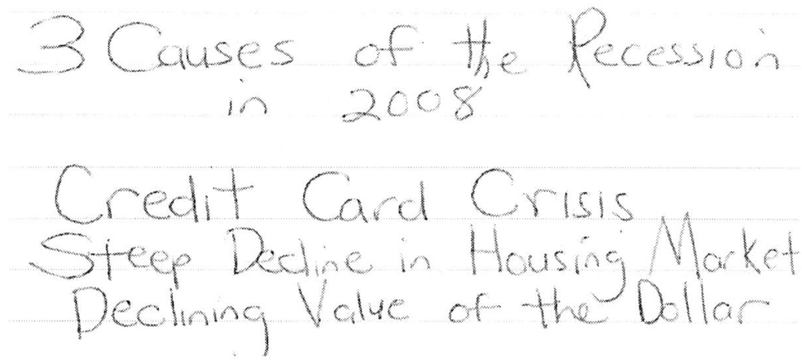

Figure 2. Example of a journal entry.

Table 4. Sample format for journal entry

Great Depression Causes	2008 Recession Causes
1. 2. 3.	1. 2. 3.
Great Depression Effects	2008 Recession Effects
1. 2.	1. 2.
Great Depression Era Government Action	2008 Government Action Taken
1. 2.	1. 2.
Great Depression Era Government Reform	Future Government Reform
1. 2.	1. 2.

DAY 5

Evaluation

On Day 5, each team will be given five to eight minutes to present their projects to the class. Examples of two slides from two different student presentations are below (see Figure 3). The students will be graded based on the points outlined in the summative rubric (see Appendix C).

Figure 3. Examples from actual PowerPoint presentations.

Extension

Depending on school and district resources, a field trip to the city hall could be arranged to examine the city budget and talk to the comptroller of a town to determine the status of the budget. City or town officials could be interviewed to determine how taxpayer dollars are spent, how the budget is balanced, if there are issues that are causing a crisis within the budget, and how these matters are handled locally.

Additionally, perhaps the mayor's office could be contacted, and students could be invited to learn how the city's budget is created and prioritized. In addition to sharing information about the city's budget, the mayor or his representative could invite students to share their views of how the city should prioritize the budget during these challenging economic times.

In some cities, the mayor and city council members from the city's budget office present budget workshops. Students could attend to see firsthand how these procedures are carried out at the local level; the public is generally invited to attend any of the budget workshops. Students could see how city officials present information and analyze the budget. Residents are also given an opportunity to give input to city leaders about their budget priorities during these workshops.

If school funds are limited, perhaps one of these city officials or a university economics professor could be invited as a class guest. Students could interview these guests as part of a panel on city budgets.

Students could create a comic strip related to the process of a great depression or recession (such as the stock market does this, then the lenders/banks do this….)

Students could create a job fair, but instead of getting jobs at corporations, the students posing as potential employers could be a failed bank, a bad stock, and a realtor company trying to sell foreclosed homes. This shows the potential "employee" that it would be hard to live and operate in a job market or economy where your only choices are failed banks, and a bad stock or bad investment firm.

APPENDIX A

Participation and Cooperation Rubric

	16	*12*	*8*	*4*	*Score*
Cooperation	Students worked extremely well together, demonstrating extensive time and effort. Accomplished all goals. Shared responsibilities and compromised when needed.	Students worked well together. Compromised when needed. Shared responsibilities, time and effort equally.	Some cooperation problems arose. Students did not always compromise well. All did not share responsibilities, time, and effort equally.	Many problems within group.	
	Students used extremely supportive language and encouraged each other.	Students used supportive language, encouraged each other.	Some students were supportive, occasionally using supportive language and providing some level of encouragement.	Evidence of competitiveness among group members; use of negative motivation or encouragement.	
Engagement	Student proactively contributed to class by offering ideas and asking questions more than once per class.	Student proactively contributed to class by offering ideas and asking questions once per class.	Student rarely contributed to class by offering ideas and asking questions.	Student never contributed to class by offering ideas and asking questions.	

APPENDIX B

Checklist for Journals

Great Depression causes	1 2 3
2008 Recession causes	1 2 3
Great Depression effects	1 2
2008 Recession effects	1 2
Comparison of health care accessibility	1 2
Great Depression Era government action	1 2
2008 Recession government action taken	1 2
Great Depression Era government reform	1 2
Future government reform	1 2
	/20

APPENDIX C

Summative Rubric

Category	*A*	*B*	*C*	*F*
Accuracy	(21–25) All information is accurate/gives author credit	(17–20) Most information is accurate/gives author credit	(13–16) Some information is accurate/gives author credit	(0–12) No project/poor info/irrelevant info/lack of credit to author
Effort	(21–25) Completed project turned in/pictures, color adds to the project	(17–20) Project mostly done with pictures and color	(13–16) Some of the project done with a limited number of pictures	(0–12) No project/poor effort, color scheme/pictures take away from project
Completion	(17–20) Project has 12-15 slides	(13–16) Project has 9-11 slides	(8–12) Project has 5-8 slides	(0–7) Project has fewer than 5 slides or not done
Creativity	(17–20) Presidential recommendations are relevant and supported by information	(13–16) Presidential recommendations are misguided or not completely supported by facts	(8–12) Presidential recommendations are random and/or incomplete	(0–7) There are no presidential recommendations
Presentation	(10) Presentation is 5-8 minutes	(8–9) Presentation is too short or long by 1-2 min	(6–7) Presentation is too long or short by 2-3 min	(0–5) Presentation is too long or short by 3+ min or not done at all

William Paul
Economics Teacher,
Emmett J. Conrad High School, Dallas ISD

Mary Margaret Capraro
Teaching, Learning, and Culture,
Texas A&M University

JENNIFER STITT AND SERKAN ÖZEL

14. GEOGRAPHICAL INFORMATION SYSTEMS (GIS): DEFINE THE WOLRD AS YOU SEE IT

What does the world look like to you?

WELL-DEFINED OUTCOME

Students will use what they know about the World Geography and learn about topology to depict landmasses and oceans on a scale model of the Earth. They will use geometry to construct the model and to determine its volume and surface area.

INTRODUCTION/BACKGROUND

Students will create a globe to scale out of balloons and papier-mâché and then use pictures and words from magazines and other media to represent their perspectives about different continents. The project will focus on the outlets and environments that shape ways of thinking about the world. This project can be given as the first project of the year in Geographical Information Systems (GIS) class, prior to teaching the software, to help students develop classroom behaviors related to the use of computers in classrooms. Students do not always enjoy sitting at the computer, and so this project involves them in a hands-on experience. The idea is for students to blow up balloons and then papier-mâché them. We coated the balloons twice. Then the students paint the balloons blue to represent water. At this point, we discussed how much of the earth is covered by water. After the balloons are painted, the students draw in the continents. The continents need to be drawn to scale. I[1] did not allow the students to print out continents via a webpage and glue them to the balloon; I asked them to attempt to draw the shapes themselves. Thus, Africa cannot be smaller than North America. Inside the continents, the students will collage the pictures that they collected from magazines and other media to fit inside the outline of the continent. They cannot place pictures outside the boundaries of the continents; the shapes of the continents should be visible. For example, in North America, students may place a picture of a hamburger and the current president of the United States. Students may use words and/or pictures. Students will also measure surface area and volume. At the end of the project, students will present their globe to class and can present their reasoning for their selection of pictures in a PowerPoint presentation.

At the conclusion of this project, students will have built a connection to the earth and the world they live in. Student engagement in computer software related to the earth has increased, if students first built an interest in the earth and their surroundings. Hands-on projects help students connect with other students in the class as well as connect with the concepts learned. In addition, a hands-on approach to circumference and surface area helps the students visually connect this mathematical concept to the concepts related to the earth and the world.

OBJECTIVES

Mathematics

- The student uses geometric thinking to understand mathematical concepts and relationships. Spatial reasoning plays a critical role in geometry. Geometric figures provide powerful ways to represent mathematical situations and to express generalizations about space and spatial relationships.
- Geometry involves the study of geometric figures of zero, one, two, and three dimensions and the relationships among them. The student is expected to understand the properties and relationships related to the size, shape, location, direction, and orientation of these figures.
- The student analyzes geometric relationships in order to make and verify conjectures.

R. M. Capraro, M. M. Capraro, J. Morgan and J. Scheurich, (eds.), A Companion to Interdisciplinary STEM Project-Based Learning: For Teachers by Teachers, 127–132.

- The student is expected to use constructions to explore attributes of geometric figures and make conjectures about geometric relationships.
- The student uses tools to determine measurements of geometric figures and extends measurement concepts to find perimeter, area, and volume in problem situations.
- The student finds areas of regular polygons, circles, and composite figures.
- The student uses numeric and geometric patterns to make generalizations about geometric properties, including properties of polygons, ratios in similar figures and solids, and angle relationships in polygons and circles.

Social Science

- The student uses geographic tools to collect, analyze, and interpret data.
- The student is expected to create thematic maps, graphs, charts, models, and databases representing various aspects of the United States and pose and answer questions about geographic distributions and patterns shown on maps, graphs, charts, models, and databases.
- The student understands how physical and human characteristics of place can change.

Language Arts

- The student is expected to use prewriting strategies to generate ideas, develop voice, and plan.
- The student is expected to compile written ideas and representations into reports, summaries, or other formats and draw conclusions.

GIS Elective/Tech Applications

- The student is expected to gain problem solving skills and group work skills by seeking and responding to his or her peers and professionals in delineating technological tasks and problem solving.
- The student is expected to develop projects according to defined output specifications including target audience and viewing environments.
- The student is expected to assess how people's changing perceptions of geographic features have led to changes in human societies.
- The student understands how to use geographic knowledge, skills, and perspectives to analyze problems and make decisions.
- The student understands how spatial features influence human behavior.
- The student knows how to use technology to represent and interpret the Earth's physical and human systems.

MATERIALS NEEDED

- Balloons (regular size)
- Newspaper
- Water
- Flour
- Spoons (to stir the mixture)
- Tupperware bowls (for the water and flour mixture)
- Magazines
- Regular glue
- Mod Podge glue
- Blue Tempera paint
- PowerPoint (PC) or Keynote (Mac)
- Printer
- Pens/pencils
- Scissors
- Internet access
- String
- Rulers

STUDENT INTRODUCTION

What does the world look like through your eyes? If you were to cut pictures and words from magazines, what pictures and words would you choose? Do you see the world as mostly negative, positive, or both? Are there certain celebrities that have shaped the way you see the world? Do you think the world is changing both politically and environmentally? What has the media done to skew your thinking of the world? You will create a globe, and you will use words and/or pictures to represent your view of the world. However, these pictures and/or words must fit entirely inside the shapes of the continents. Therefore, your project will look like an actual globe, but instead of land features, you will use pictures and/or words to represent how you see the world. Important concepts in the lesson include scale, surface area, and volume. An item will be placed in the room to aid you in your measurement of these values. You will have to find the object somewhere around the room to help you measure. You will also need a mathematical formula chart to solve for surface area and volume.

ENGAGEMENT

In order to gain interest in the project, the teacher starts off the project by having an open discussion about how environment shapes the way people see the world. The teacher, then, starts a dialogue with his or her students and has them discuss how they would define the United States using only pictures and/or phrases. It is most likely that not all students know all seven continents. A discussion about Antarctica can be raised because it is common that students do not know about Antarctica. It is important that this discussion does not focus on gathering information about the continents but on sharing students' opinions of the continents. However, students can be provided an opportunity to do Internet research about the continents. If students do not have an opinion, the teacher can have the students conduct research and ask them to start forming an opinion.

EXPLORATION

After the engagement period, the teacher presents student handouts and has students brainstorm on paper how they feel about the different parts of the world. The teacher can continue the discussion about what pictures and words come to students' minds when they think of each continent. Students may work by themselves or in pairs. Once students have brainstormed and have an idea, they may start the creation of the globe.

EXPLANATION

After the students have a direction and have grasped their opinions about the world, each student or pair will get one balloon, a bowl, one to two cups of water, one cup of flour, and some newspaper. They are to add the flour to the water slowly and stir as they pour. One person should add the flour while the other person mixes. If teachers prefer to create a premade mixture prior to class, they may do so. The water to flour mixture should be a 2:1 ratio (this step provides a good opportunity to discuss chemistry and ratio). The teacher explains to the students that they always can add more flour or water later; the teacher should emphasize that students can start out with a small amount and work from there. After the mixture is made, the students need to start tearing the newspaper lengthwise into narrow strips. Students will then dip the strips into the mixture. As the students pull out the strips, they should wring them out with their fingers to knock off any excess mix. Students are to cover their balloons with a thin coat of papier-mâché. Some students will be slow and detailed and other students will finish early. The coating process can extend to another lesson. Let the first coat of papier-mâché dry overnight. The students should put their names on a piece of paper and set the balloon on top of the paper so that the teacher and students will be able to distinguish the balloons. Students need to wash out their bowls and spoons so they are clean for the next day. Cleaning time should be incorporated into the schedule.

The following day, students will coat their balloons with another layer of papier-mâché. Even though two coats should be adequate, students can layer on three or four coats as they wish. The teacher should be aware that coating the balloon and cleanup should not take the entire 90 minutes. Students can start looking at magazines to get ideas of what pictures they might use. Students should not cut out the pictures until they have drawn their continents because the pictures need to fit entirely inside the shapes they draw.

After the two coats have dried, students will use blue tempera paint to paint their entire globe blue. They may want to do a least two coats of paint or at least enough to keep the newspaper from showing through. After the paint has dried, the students will begin to draw their continents to scale. Some students will have a difficult time with this, and they will want to paste pictures from the Internet onto their globe. A teacher may let his or her students use pictures from the Web. However, encouraging them to draw their own continents will increase their creativity. After the continents are drawn, the students can start cutting out pictures and/or words from magazines. The teacher and students should provide magazines because numerous magazines will be needed.

To place the pictures on the globe, the students will use Mod Podge. They will coat the back of the picture to glue it down and then paint over the top of the picture with the glue. When the glue dries, it will create a shiny surface. Students will need to cut out small pictures to fit inside their continents. The teacher may want to see the shape of the continent by the use of pictures. After all the pictures are glued down, brushing one final coat of Mod Podge over the entire globe is recommended. This cutting and placing pictures procedure takes several class periods. However, in a block schedule (1.5 hours of class time), students will be able to finish cutting and pasting the pictures. Because students will be working with paint and glue, the teacher should remind them to wear an old t-shirt.

After the globe is finished, students will need to find the surface area and volume of the globe. Students will need string and a math formula chart. With the string, students can measure circumference, and using circumference and radius/diameter, they can deduce surface area and volume.

After the globe is finished or during the globe creation, students will create a PowerPoint presentation about their globe and why they chose the pictures they did. They may have slides for each continent or just a few slides to help keep them on track. Moreover, they will present their surface area and volume results and how they did calculations. The teacher should emphasize eye contact as an important presentation technique. Students will be graded on overall look of the globe, their work ethic during its creation, and their presentation. Points should be deducted every time students present an informative message instead of an opinion (see Presentation and Globe Rubrics).

EXTENSION

As an extension for this project, students can create another globe, but instead of using pictures and words to define the world as they see it, they will use pictures and words to define the world as they would like to see it. What would make a better world? The same procedures for creating the globe can be followed.

EVALUATION

See rubrics in Appendix A and Appendix B.

NOTES

[1] "I" in this section represents *Jennifer Stitt*, who conducted this project in a real classroom setting.

APPENDIX A

Presentation Rubric - 16 points

	Exemplary 4	*Mastery 3*	*Proficient 2*	*Introductory 1*
Uses visuals clearly and effectively	Highly developed, memorable visuals illustrate main point. Manages visuals well.	Visuals clarify and illustrate main points. Presentation is memorable, and visuals are neat and well organized.	Visuals clarify and illustrate main points.	Visuals are not tightly linked to presentation and do not support or clarify main points.
Communicates and stresses main points	Communication is articulate and well thought out. Main points are very clear.	Communication is good; most of the main points are covered.	Communication is somewhat lacking, but main points are still evident.	Communication is poor, and main points are not fully covered.
Body posture projects confidence and authority	Student stands tall and has confidence.	Student almost always stands tall and has confidence.	Student is nervous, but it still able to convey the project.	Student is unsure what to say and is not proud of his/her work.
Makes minimal pauses and avoids filler words	Student does not stumble over words. The speech flows perfectly. Student does not use filler words.	Student does not stumble over words, but uses a few filler words.	Student has a hard time getting the words out but eventually is able to communicate effectively.	Student consistently stumbles over words and uses filler words several times.

APPENDIX B

Globe Rubric - 16 points (counts twice = 32 points)

	Exemplary 4	*Mastery 3*	*Proficient 2*	*Introductory 1*
Required Elements	The balloon included all elements. All math problems were solved.	The balloon included all elements. Most of the math was complete.	The balloon included some elements. An attempt was made at the math.	The balloon was mostly empty and math was not present.
Graphics - Relevance	All graphics are related to the topic and make it easier to understand.	All graphics are related to the topic, and most make it easier to understand.	Not all graphics related but were easy to understand.	Graphics were hard to understand and were not related to the topic.
Attractiveness	The balloon is exceptionally attractive in terms of design, layout, and neatness.	The balloon is attractive in terms of design, layout, and neatness.	The balloon is acceptably attractive though it may be a bit messy.	The balloon is distractingly messy or very poorly designed. It is not attractive.
Participation	Student was actively engaged during the project.	Student was engaged for most of the project.	Student had to be told a few times to get back on task.	Student was off-task for most of the project.

EXAMPLES OF STUDENT WORK

Figure 1. An example of student work.

Figure 2. An example of student work.

Jennifer Stitt
GIS Teacher,
Emmett J. Conrad High School, Dallas ISD

Serkan Özel
Educational Psychology,
Texas A&M University

JENNIFER STITT AND SERKAN ÖZEL

15. GEOGRAPHICAL INFORMATION SYSTEMS (GIS): NEIGHBORHOOD PROJECT

If you could redesign your neighborhood, what it look like and what would you do?

WELL-DEFINED OUTCOME

At the conclusion of this project, students will have an understanding of composite shapes, volume, surface area, vertical angles, perpendicular lines, and parallel lines. Furthermore, they will increase their geospatial knowledge by developing a city where certain items are placed in certain areas to fit the need of the geospatial extent. In addition, students will gain an understanding of the inner workings of a city and why certain elements of a city occur. Students will also draw conclusions as to what items draw people into the city and what items push people out of certain areas of a city.

INTRODUCTION/BACKGROUND

The purpose of this project is for students to create a three-dimensional representation of their current neighborhood. However, it can be the neighborhood in the future or a redesign of the current spatial extent. This project requires a lot of hands-on time, and students will design a brochure to entice people to revisit the area they designed. Students work in groups of four students. Groups will assign each of their members with specific roles; for example, each group will have a project manager. The project manager is responsible for overseeing the project and project materials. The planner is the second important role in each group. The planner is responsible for keeping consistency in the project and making sure that each section of the neighborhood fits with the overall theme of the project. The groups will also need to assign someone to the role of recorder; this person will keep a log of all the new or existing stores, offices, and structures that are built. The recorder should also be able to explain the purpose of all buildings in the group's neighborhood. The final role responsibility is the parks and recreation officer who will make sure that each neighborhood preserves some green areas.

Each group will receive four pieces of particle board—one for each group member. Each student will design a part of the group's neighborhood on his or her particle board. However, group members have to communicate with each other about how to design their parts in order to keep coherence and consistency in the neighborhood. In the sample project I[1] conducted in my classroom, I kept our school as a reference point for neighborhoods that students were redesigning. Our school could be on the center of four particle boards or on any particle board. Because I had access to a plotter, I could zoom in on the students' portion of their maps and print them out as street level maps to the size of their four pieces of particle board. If students design the neighborhood of the future, they can demolish streets and/or add new streets.

This project can be extended to a city-scale redesign project; students can either redesign the city they live in or design a fictitious city they would like to live in. However, I preferred having students redesign their own neighborhoods because most of my students were from lower income areas, and I wanted them to take ownership of their neighborhood. When I first conducted this project, I let students create buildings that fit into their neighborhoods. However, it would be helpful to have some requirements about keeping the shapes of buildings such as cylinders, pyramids, rectangular prisms, and so on. This would provide an extension to geometry topics like area, volume, parallel lines, vertical angles, slope, and so on. After students finish constructing the three-dimensional design of their neighborhood, they will work collectively to create a brochure of their area to distribute to prospective residents.

R. M. Capraro, M. M. Capraro, J. Morgan and J. Scheurich, (eds.), A Companion to Interdisciplinary STEM Project-Based Learning: For Teachers by Teachers, 133–138.

OBJECTIVES

Mathematics

- The student is expected to develop spatial reasoning.
- The student is expected to use geometric figures to represent mathematical situations and to express generalizations about space and spatial relationships.
- The student is expected to use geometric thinking to understand mathematical concepts and relationships.
- The student is expected to explain properties and relationships having to do with size, shape, location, direction, and orientation of these figures.
- The student is expected to analyze geometric relationships in order to make and verify conjectures.
- The student is expected to be able to use constructions to explore attributes of geometric figures and to make conjectures about geometric relationships.
- The student is expected be able to use tools to determine measurements of geometric figures and extend measurement concepts to find perimeter, area, and volume in problem situations.
- The student is expected to find areas of regular polygons, circles, and composite figures.
- The student is expected to be able to use numeric and geometric patterns to make generalizations about geometric properties, including properties of polygons, ratios in similar figures and solids, and angle relationships in polygons and circles.

Social Science

- The student is expected to understand the influence of scientific discoveries and technological innovations on daily life in the United States.
- The student is expected to be able to use problem-solving and decision-making skills when working independently and with others in a variety of settings.
- The student is expected to use a problem-solving process to identify a problem, gather information, list and consider options, consider advantages and disadvantages, choose and implement a solution, and evaluate the effectiveness of the solution.
- The student is expected to use a decision-making process to identify a situation that requires a decision, gather information, identify options, predict consequences, and take action to implement a decision.
- The student is expected to identify the source, use, quality, and conservation of water.
- The student is expected to document the use and conservation of both renewable and non-renewable resources.

Language Arts

- The student is expected to use prewriting strategies to generate ideas, develop voice, and plan.
- The student is expected to develop drafts both alone and collaboratively by organizing and reorganizing content and by refining style to suit occasion, audience, and purpose.
- The student is expected to proofread writing for appropriateness of organization, content, style, and conventions.
- The student is expected to refine selected pieces frequently in order to publish them for general and specific audiences.
- The student is expected to use technology for aspects of creating, revising, editing, and publishing texts.

Science

- The student is expected to make responsible choices in selecting everyday products and services using scientific information.
- The student is expected to evaluate the impact of research on society, scientific thought, and the environment.
- The student is expected to make wise choices in the use and conservation of resources and the disposal or recycling of materials.

GIS Elective/Technology Applications

- The student is expected to know and understand contemporary issues in the context of spatial and environmental perspectives.
- The student is expected to identify renewable and non-renewable resources that must come from outside an ecosystem such as food, water, lumber, and energy.
- The student is expected to analyze and evaluate the economic significance and interdependence of components of the environmental system.

MATERIALS NEEDED

- Scissors
- Hot glue guns
- Glue sticks (large amount)
- Glue
- Duct tape
- String
- Computers with Internet access
- Microsoft Word
- Microsoft PowerPoint
- Rulers
- Printer/plotter (42 by 150 inch)
- Math chart (volume, circumference, etc.)
- Card stock
- Construction paper
- Pipe cleaners
- Metal wire
- Christmas lights
- Paint (black, green, brown, silver, and blue were the top colors used)
- Tin Cutters (for cans; they work better than scissors or X-acto knives)
- Recyclable materials (anything from around the house that might work)

STUDENT INTRODUCTION

You have been given an endless amount of money! However, you need to use that money to rebuild your neighborhood. What would you add; what would you take away? You will use your school as an anchor for your design. It will also serve as a focal point for your neighborhood redesign project. You are only limited by your imagination!

You may pick three other city employees to help you build and/or redesign your neighborhood. Each of you needs to pick a role to play in the group. The roles are project manager, recorder, planner, and parks and recreation officer. The project manager will oversee the entire project and be responsible for communicating to other group members. The recorder will write down every structure that was torn down and every structure that was built. The recorder will also be responsible for keeping track of any values measured for surface area and volume. The planner helps keep the overall theme in place and makes sure the group doesn't have too much of one item and not enough of other items. Lastly, the parks and recreation officer will be responsible for green areas in the neighborhood. Unless the group can justify a reason for no green spaces, concrete-jungle neighborhoods will not be accepted. Even though everyone has their own role to play and piece to design, the four parts in each group will build the neighborhood. Thus, consistent scaling is very important. For example, the buildings must complement each other in size. Tell the students that more details will be coming soon!

ENGAGEMENT

Students start the project by exploring websites about three-dimensional city models. The teacher gives handouts that include websites for students to investigate, such as:

- http://www.zmapping.co.uk/#
- http://www.turbosquid.com/search/city/?gclid=CJ-qjeaE4pkCFSQeDQodaikyXA

- http://www.global3darts.com/
- http://www.the3dstudio.com/
- Google : 3D City Models

After students have spent some time exploring the websites, the teacher initiates a discussion to elicit students' perspectives about the designs they saw on the websites. After the discussion is over, the teacher provides students with handouts and lets students form their groups. However, the teacher may prefer forming the groups rather than allowing students to choose their group members. Based on the amount of class time remaining, the teacher can guide students to a website that contains composite shapes (e.g., http://www.senteacher.org/Worksheet/12/Nets.xhtml). Students can use such composite shapes to design their buildings.

EXPLORATION

After students have grasped the concept of three-dimensional cities and composite shapes, the teacher allows students to sit with their group and brainstorm what items they think their neighborhood needs. The teacher and students also work on getting overlay maps printed to fit over students' boards. Students may choose a customary outline form to begin brainstorming so they will have direction when they begin their project. Students need to discuss the types of items their neighborhood already has. Students may want to start making a list of stores, schools, and other facilities that will be part of their redesign project. At this point in the project, group members will decide their roles. The teacher should give students about thirty minutes to discuss the roles and their assignment. After the discussion is over and students have their own group role, students need to gather their overlay maps. If the plotter access is limited, or there is no plotter at all, students can draw their own map. Students are free to choose the part of their neighborhood they want to redesign; thus, they choose their zoom level on the map.

EXPLANATION

Students have been presented with the idea that they are in charge of redesigning their neighborhood. The student handout will be given and students will pick their groups. After they divide into groups of four, students will each be handed a thin piece of plywood or cardboard that is two square feet, and they will write their names on the back of their pieces. The teacher should prepare these pieces of plywood or cardboard before class.

If a large-format printer is available, the teacher prints a map from either the Internet or GIS software that represents the area in which the school is located. The teacher asks the students to position the school in any quadrant they want and then make the map accordingly. The students can assist in the spatial extent process. If a large-format printer is not available, students can look at smaller maps and estimate the location of streets. The teacher should ask students to redesign their neighborhoods without changing current streets.

EXTENSION

As an extension to this project, students can draft a city-planning guide as to how their city will be run and what important aspects will govern the city. What departments will their city have? The students should describe their importance. Furthermore, are there future city projects that need funding, and how will they obtain this funding?

EVALUATION

Rubrics found in Appendix A and B can be used to assess the project.

NOTES

[1] "I" in this section represents *Jennifer Stitt*, who conducted this project in a real classroom setting.

EXAMPLES OF STUDENT WORK

Figure 1. An example of student work.

Figure 2. An example of student work.

APPENDIX A

Presentation Rubric □ 16 points

	Exemplary 4	*Mastery 3*	*Proficient 2*	*Introductory 1*
Use of Visuals	Highly developed, memorable visuals illustrate main point. Manages visuals well.	Visuals clarify and illustrate main points. Presentation is memorable, and visuals are neat and well organized.	Visuals clarify and illustrate main points.	Visuals are not tightly linked to presentation and do not support or clarify main points.
Communication of Main Point	Communication is articulate and well thought-out. Main points are very clear.	Communication is good; most of the main points are covered.	Communication is somewhat lacking, but main points are still evident.	Communication is poor, and main points are not fully covered.
Body Posture	Student stands tall and has confidence.	Student almost always stands tall and has confidence.	Student is nervous but is still able to convey the project.	Student is unsure what to say and is not proud of his/her work.
Fluency	Student does not stumble over words. The speech flows perfectly. Student does not use filler words.	Student does not stumble over words, but uses a few filler words.	Student has a hard time getting the words out but eventually is able to communicate effectively.	Student consistently stumbles over words and uses filler words several times.

APPENDIX B

Redesign Model □ 30 points

	Exemplary 4	*Mastery 3*	*Proficient 2*	*Introductory 1*
Inclusion of Required Elements	Student included a well thought-out plan and brochure.	Student included a mediocre plan and brochure.	Student included some of the required elements but not all.	Student included one if any of the required elements of the plan and brochure.
Overall Look	Student typed the brochure and included visual images from the redesign project.	Student typed the brochure but did not include visual images from the redesign project.	Student typed the brochure but missed some information.	Student handwrote the brochure.
Originality	Student was able to come up with a redesign plan that was thought-provoking and original. Student did not copy from another student. Student thought outside of the box and did not try to just "get it done."	Student came up with a semi-interesting plan, and the redesign project was ordinary.	Student came up with a plan, but it seemed to lack something and did not have direction.	Student could not come up with a plan and had to ask the teacher several times what to do. Student often copied from what other students were doing.

Jennifer Stitt
GIS Teacher,
Emmett J. Conrad High School, Dallas ISD

Serkan Özel
Educational Psychology,
Texas A&M University

M. SENCER CORLU AND KATHY VONDRACEK

16. PLAN YOUR INTERNATIONAL VACATION

How can students turn their vacation into a meaningful experience?

WELL-DEFINED OUTCOME

Students using internet resources will identify and explain the migration influences of places and regions based on the travels of Marco Polo, and identify and explain the geographic factors responsible for the location of economic activities in places and regions as they convert currencies and work out a budget in planning their international vacation to at least 5 countries on the Silk Road.

INTRODUCTION

Marco Polo (1254–1324) is probably the most famous traveller of all times. His journey on the Silk Road through Asia lasted 24 years. He travelled the whole of China and returned to tell the tale, which became the greatest travelogue. Following the steps of Marco Polo, students will plan their own travel hitting the road in Dallas, fly to Europe then go all the way to China. They will experience different cultures and customs on their way, anticipate the difficulties they may experience in advance, and be ready to use the skills they learned in school to overcome those difficulties. Their plans will not be limited to the available features of the free Google Earth software, but will include information about geological particularities of a country of their choice, as well as a detailed budget for their travel. This PBL follows an interdisciplinary approach, and will enable students to see links among subjects such as chemistry, geology, history, and mathematics.

OBJECTIVES

Science

- Students will understand that science is a way of learning about the natural world.
- Students should know how science has built a vast body of changing and increasing knowledge described by physical, mathematical, and conceptual models, and also should know that science may not answer all questions.
- Students should be able to express and manipulate chemical quantities using scientific conventions and mathematical procedures such as dimensional analysis, scientific notation, and significant figures.
- The student uses critical thinking, scientific reasoning, and problem solving to make informed decisions within and outside the classroom.
- The student will research and describe the history of science and contributions of scientists.
- The student will research and evaluate measures to minimize harmful environmental factors on body systems.

Environmental Sciences

- The student knows the relationship between carrying capacity and changes in populations and ecosystems.
- The student is expected to summarize forms and sources of energy and analyse and make predictions about the impact on populations of geographic locales, natural events, diseases, and birth and death rates.

Social Science and History

- The student understands the characteristics and relative locations of major historical and contemporary societies, locate major historical and contemporary societies on maps and globes, identify and explain the geographic factors responsible for patterns of population in places and regions, explain ways in which human migration influences the character of places and regions, and identify and explain the geographic factors responsible for the location of economic activities in places and regions.

R. M. Capraro, M. M. Capraro, J. Morgan and J. Scheurich, (eds.), A Companion to Interdisciplinary STEM Project-Based Learning: For Teachers by Teachers, 139–144.

- English
- The student organizes his/her ideas in writing to ensure coherence, logical progression, and support for ideas, uses prewriting strategies to generate ideas, develop voice, and plan.
- The student is expected to compile written ideas and representations into reports, summaries, or other formats and draw conclusions.

Mathematics

Number, operation, and quantitative reasoning.
- The student understands that different forms of numbers are appropriate for different situations.
- The student is expected to express numbers in scientific notation, including negative exponents, in appropriate problem situations.

Patterns, relationships, and algebraic thinking.
- The student uses graphs, tables, and algebraic representations to make predictions and solve problems.
- The student is expected to predict, find, and justify solutions to application problems using appropriate tables, graphs, and algebraic equations.

Measurement.
- The student uses indirect measurement to solve problems.
- The student is expected to use the Pythagorean Theorem to solve real-life problems.
- The student describes how changes in dimensions affect linear, area, and volume measures.

MATERIALS NEEDED

http://www.silkroadfoundation.org/artl/chrono.shtml

STUDENT INTRODUCTION

Long before Christopher Columbus left Europe to find an alternative way to India, but ended up being in America, many other travellers left Europe to see what the East is like by travelling on the traditional route, what was then called the *Silk Road.* You will investigate why this path is called the Silk Road; learn more about the infamous traveller *Marco Polo.* Marco Polo is the most famous of all western travellers who went all the way from Venice-Italy to northern China with the hope of seeing the richness and the glory of eastern civilizations at that time. We are today inviting you to join our team on our way from Dallas to Beijing by following the footsteps of Marco Polo on Silk Road. We will see whether the countries in Asia are still rich and developed as they were in the past. We will learn a lot about many of the countries on our way, including the health conditions and medical traditions of each culture. However, we will need to plan our trip very carefully before we are ready to go. We must make sure that we carry with us any necessary health products or obtain vaccinations before we leave to ensure that we maintain our health throughout the trip.

Google Earth is a great tool that can help us plan our long trip. You can see every inch of the Earth from a birds-eye point and be able to fly from one destination to another in seconds. You can use the ruler tool to calculate the distance or to see how a particular place was like 50 years ago. It also automatically calculates the best possible path between any two places in the world. Google Maps is another similar application available at http://maps.google.com. You can also use many other Internet references if you like to find information about a particular country of your choice on Silk Road.

You will be given a rubric to assess your progress for each step, and another rubric for your final output. Don't worry, although not easy, you don't have to travel on foot as Marco Polo did 1000 years ago.

Good luck!

ENGAGEMENT

Teacher starts off the project by having an open discussion about what countries students would like to see and why they would like to go and see a different country. Teacher directs students about the importance of planning a trip in advance. The teacher initiates a brainstorm session about the reasons that lead Christopher Columbus to go all the way from Europe to America. The teacher introduces Marco Polo and the Silk Road by showing students a map of one of the Silk Road paths on the overhead. Maps and a chronology of important events are available at http://www.silkroadfoundation.org/artl/chrono.shtml.

Start another discussion about the countries on the Silk Road that students may have heard before including: Italy, Greece, Turkey, Iran, Afghanistan, Pakistan, India, and China. (1–2 class period)

EXPLORATION

The next time you meet with the students; present the story that is above in the student introduction area. Next, have them explore on the Internet "Marco Polo and Silk Road". Their efforts and their presentation are going to be evaluated according to rubric 1 (see Appendix). (1 class period)

Introduce Google Earth and the available features. See Figure 1. Ruler function on top pane can be used to calculate the distance between any places on the map. Ask students to calculate the distance between their hometown and Dallas, record the distance in miles and kilometres; then find the conversion formula between miles and kilometres. (1 class period)

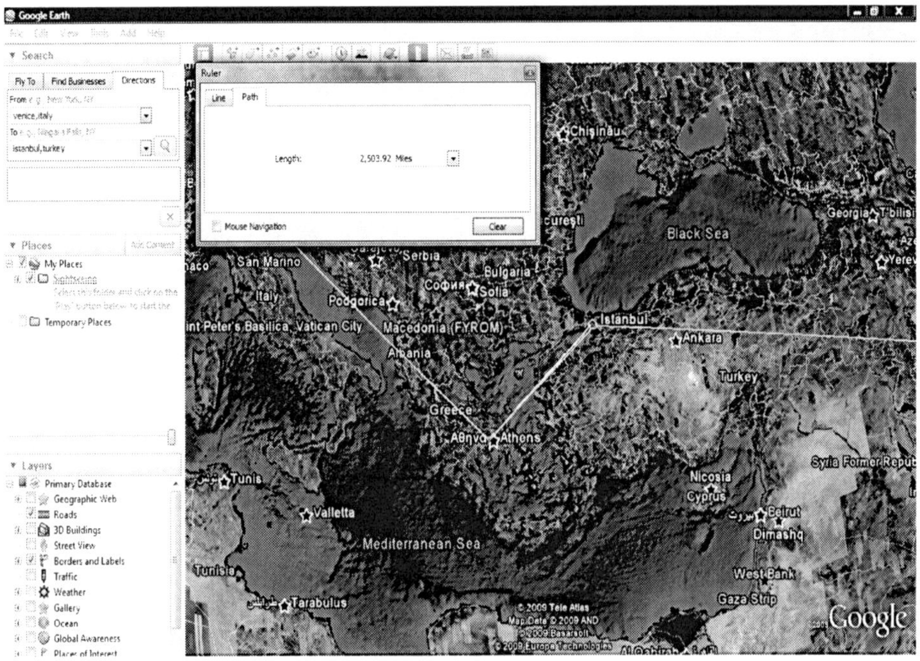

Figure 1. Sample Google Earth screenshot.

EXPLANATION

Budget planning will be the main component of this PBL. Students' final outcomes will be different from one another since they may choose to fly, drive or walk in short distances over their course of travel. There will be some constraints, though; such as

- They will have 40 days to arrive at their final destination.
- They will have to see at least 5 cities on their way to China from Dallas.

Students will use Google Earth to calculate the distance between two destinations on their Silk Road path. They are free to choose any medium of transportation, but have to consider the cost of living, travel and any health concerns in that specific country when they prepare their budget. Varying gas prices and living conditions will be a challenge for the students. They will present their budget in local currency and US dollars, and also present the exchange rates. https://www.cia.gov/library/publications/the-world-factbook/ can be an ideal resource to gather most of the information they will need. Rubric 2 (see Appendix) will be used to assess student performance. (2 periods)

The final outcome of this PBL will be a detailed presentation of a country of students' choice. Rubric 3 (see Appendix) will be used to assess their presentation. (2 periods)

EXTENSION

Students can be asked to produce a map showing all their travels and brief information about other cities they visited. Figure 2 can be used as a guide. Students can also compare the health and medical concerns of today with those that Marco Polo must have encountered on his trip. Attention should be given to any preventative methods that were used as well as prevalent diseases and their treatment.

Figure 2. A blank map of Eurasia.

APPENDIX

Rubric 1 – Research on Marco Polo and Silk Road

	RESEARCH	*PRESENTATION*
4–3	Thorough investigation of the Silk Road and the life story of Marco Polo, which includes the reasons behind his travel and what happened during his trip. A variety of resources are used, and they are indicated separately.	Great visualization supported by maps and figures.
2–1	Mediocre research of Marco Polo and his travel on the silk Road. Cited resources are not indicated clearly.	Some visualization or unclear presentation of the topics discussed in the research.
0	No attempt or mostly copy-paste work.	No attempt or totally careless presentation.

Rubric 2 – Budget Planning

	Budget
4–3	The budget includes a detailed analysis of foreseen costs of travel in both the local currency and USD. Conversion rates are available, and there is evidence that specific conditions of each country are taken into account; such as gas prices, cost of living, health concerns, etc.
2–1	A reasonable budget is prepared however, it is either not detailed, or some elements are missing.
0	No attempt or mostly copy-paste work.

Rubric 3 – Country presentation

	Budget
4–3	A detailed analysis of demographic and economic information of the chosen country is presented with emphasis on natural resource. One major natural resource is investigated in detailed, its importance and use is explained in a scientific language. The health and medical status of the country is examined and coherently explained.
2–1	A reasonable analysis of one country however, it is either not detailed, or some elements are missing.
0	No attempt or mostly copy-paste work.

Figure 3. Sample student work.

M. Sencer Corlu
Aggie STEM Center,
Texas A&M University

Kathy Vondracek
Chemistry Teacher,
Emmett J. Conrad High School, Dallas ISD

JENNIFER STITT AND SERKAN ÖZEL

17. GEOGRAPHICAL INFORMATION SYSTEMS AND BUSINESS MODELS

Why are demographic data and proximity analysis so important in developing a new business?

WELL-DEFINED OUTCOME

At the completion of this project, students will have increased their geospatial knowledge base as well as touched the surface of what defines a start-up business. In addition, students will gain a better perspective on file management, zip files, and how to download shapefiles via the Internet. Furthermore, students will develop an understanding of demographic data and how it plays into why businesses choose the locales that they do.

INTRODUCTION

GIS and Arcmap[1] are systems of software that can be used to perform business analysis on many levels. In this project, students will first explore aspects of starting a new business. Topics they will explore include mission and vision statements, business plans, obtaining capital, and marketing. Next, students will investigate the types of businesses they want to start; will their business be service oriented, sell a product, or work as a consulting firm that helps other businesses? Students will conduct research on existing business model websites about creating a business model. After they have developed their businesses and have an understanding of the type of consumer they want to attract, they will download data into Arcmap and begin a demographic and proximity analysis on where to locate their business. Students, then, will develop a map and a concrete business plan presentation. Students will have to defend or present their model to the class and explain their analyses on their choice of location for their business.

OBJECTIVES

Mathematics

- Spatial reasoning plays a critical role in geometry. Geometric figures provide powerful ways to represent mathematical situations and to express generalizations about space and spatial relationships. Students use geometric thinking to understand mathematical concepts and relationships.
- Students are expected to collect and organize data, make and interpret scatter plots, fit the graph of a function to the data, interpret the results, model, predict, and make decisions and critical judgments.

Social Science

- Students are expected to analyze political, economic, social, and demographic data to determine the level of development and standard of living in a specified area.
- Students should also be able to compare ways humans depend on, adapt to, and modify the physical environment using local, state, national, and international human activities in a variety of cultural and technological contexts.
- Students are expected to analyze economic and social data to determine the standard of living in nations.

Language Arts

- Students are expected to organize ideas in writing to ensure coherence, logical progression, and support for ideas.
- Students should be able to use prewriting strategies to generate ideas, develop voice, and plan.
- Students will compile written ideas and representations into reports, summaries, or other formats and draw conclusions.

R. M. Capraro, M. M. Capraro, J. Morgan and J. Scheurich, (eds.), A Companion to Interdisciplinary STEM Project-Based Learning: For Teachers by Teachers, 145–153.

GIS Elective/Tech Applications

- The student is expected to gain problem solving skills and group work skills by seeking and responding to his or her peers and professionals in delineating technological tasks and problem solving.
- Students will format the developed projects according to defined output specifications including target audience and viewing environments.
- Students will design and implement procedures to track trends, set timelines, and review and evaluate the products using technology tools such as GIS.
- Students will analyze demographic data to determine the level of development in a city and/or county.
- Students will develop a method to analyze place or proximity.
- Students will assess how people's changing perceptions of geographic features have led to changes in human societies.

MATERIALS NEEDED

- Computer
- GIS software (Arcmap or an equivalent)
- Printer or plotter (color preferably)

STUDENT INTRODUCTION

Marcela Lamas decides to explore the entrepreneurial voices in her head. She wants to start a business to help the community while supporting her family. Marcela first has to decide the type of business she would like to run. Will she sell a product, provide a service, or be a consultant for other businesses?

After much thought, Marcela decides to open up a local deli that has free Wi-Fi for customers and an outlet for local bands to perform. Now, she needs to obtain capital for her business, develop a business plan or model, decide on a mission statement, hire employees, and conduct demographic research on her potential customers. Then she needs to find the best location for her business based on her research.

Marcela is new in business. Thus, she is seeking suggestions from an expert in this business. She decides to consult with you about the best strategies for writing a mission statement, developing a business model, obtaining capital, and using demographic data to decide business location. She wants to discuss each step with you to benefit from your experience. She is looking forward to meeting with you in three weeks. Therefore, you need to be prepared for the meeting, and you need to print the PowerPoint of your business type, mission statement, and model/plan as well as the map showing the location of your business with information about the needs of the community and/or demographic data. Marcela may need help using Arcmap. Thus, you may need to explain to Marcela which GIS software (e.g., Arcmap) you used in order to conduct and produce your map for your business.

ENGAGEMENT

The teacher starts off the project by having an open discussion about starting a business. The teacher then leads the discussion in a way that students develop interest in starting their own businesses. The teacher facilitates brainstorming about starting a business and asks students to write down their ideas about what they would need to start a business. After students finish brainstorming and writing their ideas down, the teacher projects a list of their ideas on a screen. Then, the teacher starts a new dialogue about business types and services. This engagement section will last about a class period.

EXPLORATION

Exploration follows the engagement. To start the exploration, the teacher presents a story similar to the one provided in the student introduction section. Then, students do research on the Web about business models. They will need to take notes on what they will need in order to start a business. Also, they will need to explain how they are going to find financial support to start their businesses. This exploration section lasts about a class period.

EXPLANATION

During this class period, students should know their objectives (i.e., to help Marcela) and could have collected their data from their research on the Internet. Then they will work either individually or in pairs to create presentations about their businesses. If the students would like to work in groups, the group size should not exceed two students. First, students need to declare the business type they are going to create. Then, they need to be able to understand and explain the difference between types of businesses, such as businesses that provide a service, produce a product, or do consultancy. The students will have the flexibility to create other types of businesses.

After students adopt a business type and title their businesses, they will work on a memorable mission or vision statement for their business. The teacher should emphasize the importance of a captivating mission or vision statement and explain that the statements should be written concisely.

As a next step, the teacher requires the students to develop a business model and find a template online they can use. After the students develop their businesses, they will move into the GIS side of the project. Students should already have basic knowledge of Arcmap before they start this part of the project. If Arcmap is not available, students can create their own maps using materials such as poster board, map pencils, and markers.

First, students will need to complete an assignment related to demographic data and GIS. The assignment involves manipulating demographic data for Houston, Texas, in order to decide where to locate a cafe. To begin, the teacher should prepare the lesson "Locating an Internet Café in Houston, TX," which is provided by the Environmental Systems Research Institute (ESRI) free of charge. The specific lesson plan with data can be found at http://edcommunity.esri.com/arclessons/lesson.cfm?id=206. The teacher may access different lesson plans and their data at http://edcommunity.esri.com/arclessons/arclessons.cfm.

After the students complete the Internet Café lesson, they will need to navigate to the following census website (see http://www.esri.com/data/download/census 2000_ tigerline/index.html), choose their own state and county, and download demographic data in order to choose variables to decide where they will locate their businesses. Then, students will proceed to the "Preview and download" link. This link will take the students to the page with the United States census data (see Figure 1). In order to reach the required data, a state selection needs to be made, along with the selection of a specific county (see Figure 2).

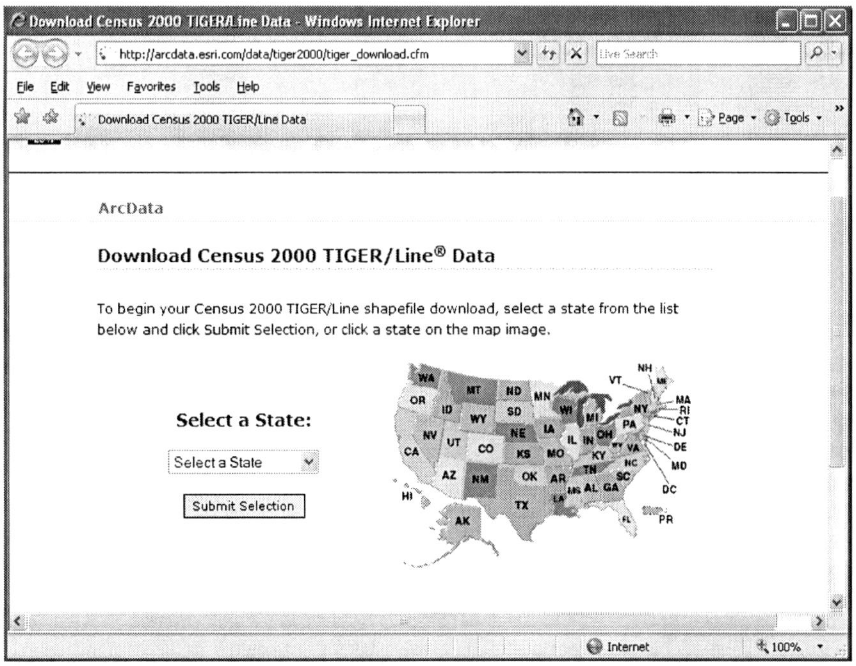

Figure 1. A screen shot of selecting a state.

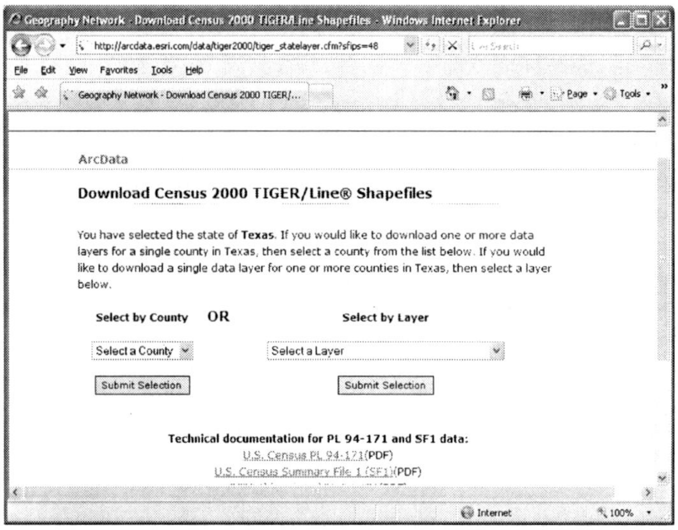

Figure 2. A screen shot of selecting a county.

Figure 3. A screen shot of data layers.

After the county is selected, the page navigates to the shapefiles page in which students are asked to select different layers for the data they will download (see Figure 3). For this specific project, students may choose from the following data:

- Block groups 2000
- Census tracts 2000
- Key geographic locations
- Line features: roads
- Urban areas 2000
- Water polygons
- Census block groups demographics (SF1)
- Census tract demographics (SF1)

This process requires several data file downloads. If there are administrative access issues on student computers, this may hinder the lesson. The teacher should take caution in advance to prevent such issues. The data files can be downloaded in advance and be provided to the students.

If students are allowed to download documents to their computers, they should select the layers and proceed to download. The shapefiles will be in zipped format (i.e., .zip files), and they will need to be extracted before they can be used. Even though zipped file icons may change across computers based on the software used, the default icon for Windows XP is presented in Figure 4.

The zipped files can be double clicked and extracted to a folder on the students' computers by clicking the "Extract all files" link. The zipped files consist of *dbf*, *shp*, and *shx* files (see Figure 5). The dbf file is a database file, and it is also called table of information, which contains demographic data. All three files together make up the shapefiles, which are either a polygon, line, or point. The shapefiles can then be imported into GIS software (Arcmap in this case). The database file will then be joined to the polygon layers. Further information about the names of layers can be accessed from the "Read Me" document located in the zipped file.

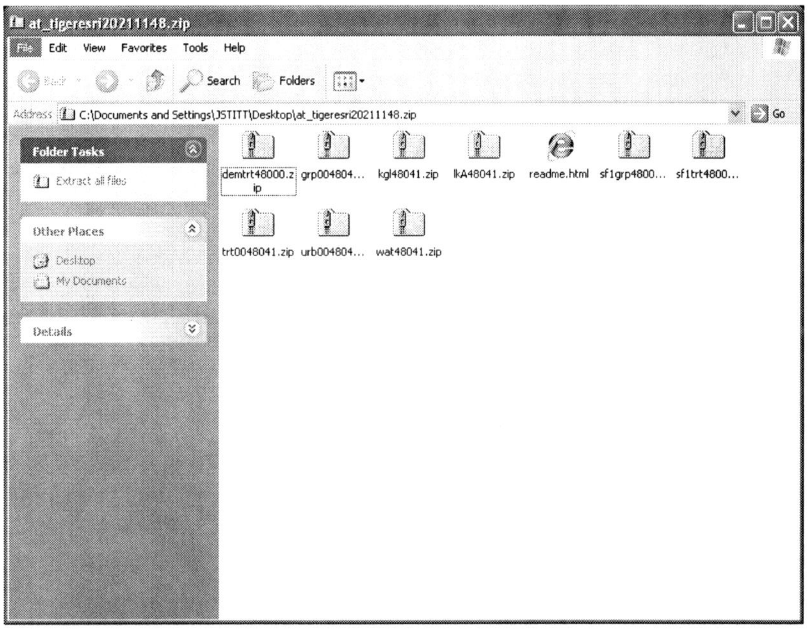

Figure 4. A screen of downloaded zip files.

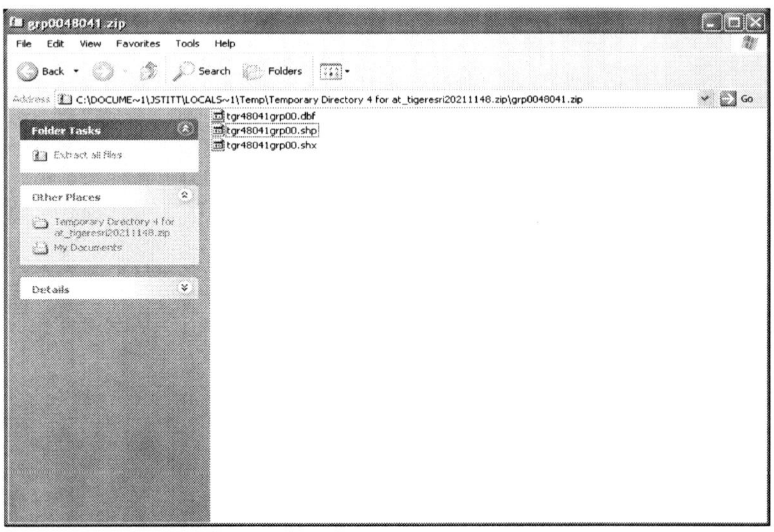

Figure 5. A screen shot of zipped files in Explorer.

Once the data is downloaded and unzipped, it can be brought into Arcmap where students can start understanding and quantifying the data. After students have worked in Arcmap on quantifying the demographic data, they need to do a proximity analysis. The questions that can lead to proximity analysis consist of "Where do the roads lead?", "Are there water sources nearby?", and "Are there any landmarks?" Moreover, age, income, and race may be important demographic factors as to where to locate the business. If students create a high-end clothing store, they should open their business in an area that would attract people in close proximity. After students have found a location, they will use the layout screen to create a map showing where their business is. Students will need to include a legend, north arrow, title, and scale in their layout (see Appendix 1). These features are built-in in Arcmap, and the teacher should be familiar with this aspect of Arcmap.

Once students have their maps ready, they will need to print them out. If students have access to a plotter, they can print their maps on large paper (e.g., 22 inches by 28 inches). Otherwise, regular printers can be used, even though the maps will print very small. Color printing will enrich the readability of the maps.

Once the map is completed, students should revisit their PowerPoint presentation. The presentation should state the name of their business, what their business does, and the type of business it is (see Appendix 2). The presentation should also include the mission and vision statement. Students must sell the class on the idea of their businesses and use concrete details about why their businesses need to be located where they are located. Students will also need to stress why they think their businesses are going to succeed. In their presentations, students must consider capital information, hiring procedures, department types, and different types of insurance for their employees. In addition, their building structure is an important issue. For example, will they need a soil analysis to start building from scratch or are they going to take over an existing structure? At the end of their presentations, students need to answer the driving question: Why are demographic data and proximity analysis so important in developing a new business? Students' participation will be evaluated using a participation rubric (see Appendix 3).

EXTENSION

As an extension, students can develop a storefront for their businesses in Google SketchUp[2] out of cardboard and paint or drawn by hand. The storefront must reflect the business they are starting. Hence, if they are starting a clothing business, they will need to have clothing pictures on their storefront.

EVALUATION

The rubrics in Appendix A, B, and C can be used to assess student knowledge of maps, their business model, and their participation in the project.

NOTES

1. Arcmap is a comprehensive software package that is for purchase. However, there might be inexpensive or free software available to the public, but it doesn't allow for in-depth analysis or data manipulation. Free software includes Google Earth and AEJEE (ArcExplorer). These software packages allow users to view and analyze some information, if Arcmap is not available.
2. Google SketchUp is 3D model software available free of charge at http://sketchup.google.com/.

APPENDIX A

Map Rubric (20 points)

	Exemplary 4	*Mastery 3*	*Proficient 2*	*Introductory 1*
Required Elements	Map contains a north arrow, legend, scale, and title. Title is catchy and layout looks exemplary. Map is zoomed to the correct extent.	North arrow, legend, scale, and title are present. Map is zoomed to the correct extent.	Something is missing: the north arrow, legend, scale, or title. Map is not zoomed to the correct location and may not be zoomed to the correct extent.	Several required elements are missing.
Labels	Elements included on the map are labeled. This includes roads, cities, states, etc. Only the elements used are the elements labeled. Labels are original and are not the default labels.	Elements included on the map are labeled. This includes roads, cities, states, etc. Only the elements used are the elements labeled.	Many elements on the map that are included are not labeled, and labels are NOT original.	Labels do not exist or do not match the elements on the map.
Graphics - Relevance	Extra graphics were used, such as inset maps, charts, or graphs, and are well placed in the layout.	Extra graphics were used, such as inset map, charts, or graphs, but they might not have been placed in the most relevant places.	An attempt to use additional graphics was made, but they did not end up on the final layout.	No additional graphics were used.
Attractiveness	The map is neat and free of any spelling errors. The legend is visible and understandable. The student went above and beyond to make the map look professional and did not add any cartoons to make it look better.	The map is neat and free of grammatical errors. Colors were used wisely.	The map is somewhat neat but confusing.	The map does not have any required elements and is not neat or professional.
Grammar	There are no grammatical or mechanical mistakes on the map.	There are 1–2 grammatical or mechanical mistakes on the map.	There are 3–4 grammatical or mechanical mistakes on the map.	There are more than 4 grammatical or mechanical mistakes on the map.

APPENDIX B

Business Model (30 points)

	Exemplary 4	*Mastery 3*	*Proficient 2*	*Introductory 1*
All elements present	Student included a well thought out vision, mission statement, objectives, strategies, and plans	Student included a mediocre vision, mission statement, objectives, strategies, and plans	Student only included some of the required elements but not all.	Student included one if any of the required elements, of vision, mission, objectives, strategies, or plans
Overall Look	Plan is typed and put in an orderly manner. Student used same chart as presented in class or developed their own chart.	Student types plan but did not use chart. However, plan is still complete and looks professional.	Student typed out a little portion of the model.	Student hand wrote the business model.
Originality	Student was able to come up with a business plan that was thought provoking and original. Student did not copy from another student. Student though out of the box and did not just try and "get it done."	Student came up with an interesting plan but that type of business is common.	Student came up with a plan, but it seemed to lack something and did not have a direction.	Student could not come up with a plan and had to ask the teacher several times what to do. Student often copied from other students what they were doing.

APPENDIX C

Participation Rubric (20 points)

	Exemplary 4	*Mastery 3*	*Proficient 2*	*Introductory 1*
Attendance	Student is present all or most days.	Student misses a few days but makes work up.	Student misses several days and does not make up all days.	Student is absent for most of the project.
Work Ethic	Student is always on-task and willing to ask questions when needed.	Student is on-task most of the time.	Teacher has to ask student more than once to stay on-task.	Student is always off-task and has to be asked numerous times to be quiet and complete the work.
Effort	Student puts in effort for most if not all of the class period. Student does not pack up and wait for the bell to ring at the door.	Student puts in a lot of effort but has been off-task once.	Student puts in little effort and is off-task several times during the class. Student packs up early.	Student is always off-task and packs up early every day.
Attitude	Student is positive.	Student is positive most of the time.	Student is not very positive but is still trying.	Student is not positive and disrupts the class.
Group Work	Student is productive when working in a group. Student actively participates.	Student is productive and only slacks off once.	Student sometimes makes other people in the group do all the work.	Student only delegates and does not work well with the group.

EXAMPLES OF STUDENT WORK

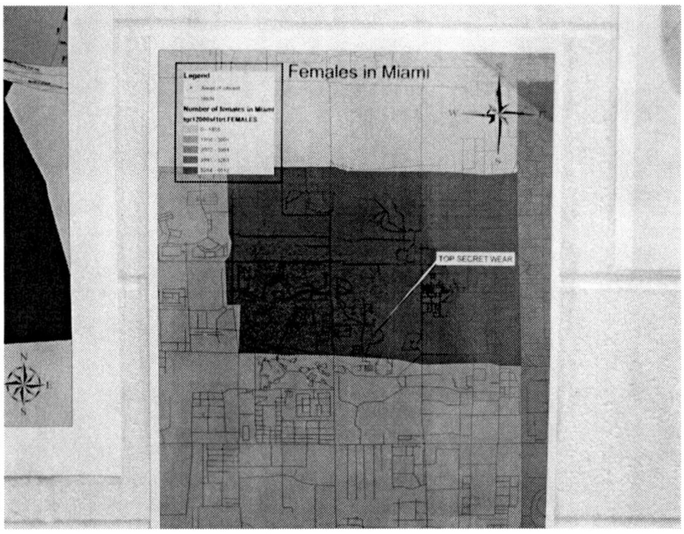

Figure 6. An example of student work: Females in Miami.

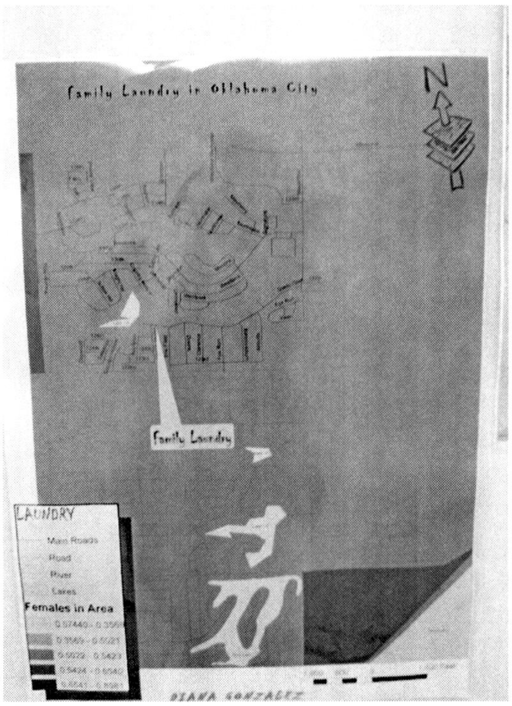

Figure 7. An example of student work: family laundry in Oklahoma City.

Jennifer Stitt
GIS Teacher,
Emmett J. Conrad High School, Dallas ISD

Serkan Özel
Educational Psychology,
Texas A&M University

JENNIFER STITT AND SERKAN ÖZEL

18. IS OUR LIBRARY FLOOR PLAN CONDUCTIVE TO LEARNING?

Student Project to Evaluate the Effectiveness of the School's Library Plan

WELL-DEFINED OUTCOME

At the conclusion of this project, students will have an understanding of higher-level AutoCAD software for designing and drawing 2-D and 3-D objects and a refined understanding of the applied geometric concepts relating 2-D to 3-D, measurements, scale, and ratio and proportion. Students will use the following geometry concepts (but may use or need to learn others) area, distance, circumference, radius, diameter, spatial reasoning, and scale factor.

INTRODUCTION

This project requires introductory knowledge of AutoCAD or equivalent two-dimensional design software. In this project, I preferred using the library due to the school's structure, but any large room with interesting angles would work. Students work in groups of four. Each person in the group is assigned a role. Then the students are given large pieces of graph paper. I printed a 32 by 32 inch, large sheet on a plotter. At the beginning of the project, students are to draft out the room/library as it currently exists. Their first draft is on the large piece of graph paper. It is up to students to pick their own scale, as discussed later. After they draft out the room on graph paper and write the dimensions, they return to the classroom and transfer their floor plan to AutoCAD or any two-dimensional design software. After the floor plan of the room is drawn in AutoCAD, the students will divide the area into equal parts. Each student in the group will be given his/her own part of the floor plan to redesign. During this process, the students have to keep in mind how they will explain why or why not certain areas of the room/library are conducive to learning. The students are only limited by their imagination, and they are expected to go beyond simply changing one wall size and a door.

OBJECTIVES

Mathematics

- The student uses geometric thinking to understand mathematical concepts and relationships. Spatial reasoning plays a critical role in geometry. Geometric figures provide powerful ways to represent mathematical situations and to express generalizations about space and spatial relationships.
- The student is expected to use geometric thinking to understand mathematical concepts and relationships.
- Students are expected to collect and organize data, make and interpret scatter plots, fit the graph of a function to the data, interpret the results, model, predict, and make decisions and critical judgments.
- The student uses inductive reasoning to formulate conjectures.
- The student uses numeric and geometric patterns to make generalizations about geometric properties, including properties of polygons, ratios in similar figures and solids, and angle relationships in polygons and circles.
- The student identifies and applies patterns from right triangles to solve problems, including special right triangles (45°-45°-90° and 30°-60°-90°) and triangles whose sides are Pythagorean triples.
- The student analyzes the relationship between three-dimensional objects and related two-dimensional representations and uses these representations to solve problems.
- The student uses one- and two-dimensional coordinate systems to represent points, lines, line segments, and figures.
- The student finds areas of sectors and arc lengths of circles using proportional reasoning.

R. M. Capraro, M. M. Capraro, J. Morgan and J. Scheurich, (eds.), A Companion to Interdisciplinary STEM Project-Based Learning: For Teachers by Teachers, 155–161.

Language Arts

- Students organize ideas in writing to ensure coherence, logical progression, and support for ideas.
- Students use prewriting strategies to generate ideas, develop voice, and plan.
- Students compile written ideas and representations into reports, summaries, or other formats and draw conclusions.
- Students write clear and effective English when preparing information.
- Students are expected to complete reports and documents that comply with the project.

AutoCAD Elective/Technology Applications

- The student is expected to be involved in decision-making and problem-solving, which are essential skills for career planning and lifelong learning.
- The student uses computer operations and applications.
- The student manages computer operations.
- The student applies the basic commands of operating-system software.
- The student employs desktop operating skills.
- The student manages file storage.
- The student applies appropriate file and disk management techniques.
- The student differentiates between files and directories.
- The student determines file organization.
- The student uses system utilities for file management.
- The student draws schematic site plans, floor plans, building elevations, sections, perspectives, and character sketches from bubble diagrams.
- The student draws scaled wall thickness plans, elevations, and sections.
- The student develops details of floor and wall sections as required.
- The student illustrates ideas for architectural projects from direct observation, experiences, and imagination.
- The student begins comparing and contrasting the use of architectural elements such as color, texture, form, line, space, and value and architectural principles such as emphasis, pattern, rhythm, balance, proportion, and unity in personal architectural projects and those of others, using vocabulary accurately.

MATERIALS NEEDED

- Tape Measure
- String
- Rulers
- Pencils
- Protractors
- Computers
- Internet access
- PowerPoint
- Plotter
- Printers
- Calculators
- Enlarged graph paper (32 inch by 32 inch) or regular graph paper
- AutoCAD software (or other two-dimensional design software)
- Cart and bin (to carry all the items and store blue prints)
- *Principles of Architecture and Construction* provides an overview of various fields of architecture, interior design, construction science, and construction technology.

STUDENT INTRODUCTION

The firm Dynamic Designs Architecture has hired you to redesign the school library. Due to modern innovation and new building standards, schools are redesigning their infrastructure in order improve the quality of life and education for students.

Your first task is to assemble a workforce. You may choose four colleagues. Responsibilities within the workforce include a project manager, drafter, and two measuring technicians. The project manager will oversee the daily progress of the group, keep the group on task, and be responsible for maintaining the drafting paperwork generated by the group. The drafter will be responsible for drafting the floor plan on enlarged graph paper. However, before drafting, the group will need to pick a factor in order to scale down the room consistently. Hence, one side of a square on the graph paper will be equivalent to a certain foot dimension in the room. Remember to pick a scale that allows the floor plan to take up most of the enlarged graph paper so the floor plan is easy to view. Finally, the two measuring technicians will be responsible for using a measuring tape to report correct measurements to the drafter.

Dynamic Designs Architecture needs an initial floor plan of the library both as a hand-drawn blueprint and in a digital format from AutoCAD. The group must submit a preliminary drawing first in order for the firm to understand and comprehend the improvements you have made. In addition, the firm will only fund the school that most needs improvement; your floor plan will help them make that decision. After receiving approval from the firm, the workforce may begin the redesign. Students need to keep in mind that the workforce's focus should be on how to make the library conducive to learning. Dynamic Designs Architecture also requires that everyone in the group payroll participate in the redesign. Good luck!

ENGAGEMENT

The teacher starts off the project by having an open discussion about what a floor plan is and by describing differences among elevations, floor plans, and three-dimensional representations. Students may look up floor plans on the Internet. Students will also form groups of four, name their group, and decide their roles in the group. Each group will need to submit a paper with their group name, group members' names, and the role of each group member. The teacher needs to emphasize that each group will be competing for a chance to redesign the library. If there is time left, the teacher will hand students a four-foot piece of string and tell them to keep the string because they might need to use it in the library[1].

EXPLORATION

The teacher begins the exploration in the classroom and hands out graph paper, measuring tape, pencils, string, rulers, and protractors to each group. Then the teacher has students walk to the library or the room of their choice as a group. Next, the teacher lets students brainstorm about how they will come up with a scale for the graph paper. Following the brainstorming, students present their procedure for determining a scale but not the scale itself.

After students have an idea about their scale, they start measuring the room without teacher guidance or direction at the beginning. After students have made several trials without teacher guidance, the teacher may provide students with hints (e.g., pick a starting point and try different measurements). The rooms may have different structures such as built-in or moveable bookcases. The teacher can initiate group discussions on solving measuring issues related to such structures. Discussions about measuring procedures, transferring collected data to graph paper, and appropriate behavior (such as working quietly in the library) are left to the teacher's timing and discretion.

EXPLANATION

After the engagement and exploration, students will meet in the library (or the selected room) every day until most of the groups are done measuring the floor plan. The students are to pick a starting point and make their way around the room, measuring the walls and built-in furniture. They also need to measure how far the built-in furniture extends beyond the wall. Students will carry scratch paper with them and record their measurements as they go. Students may draw the floor plan on graph paper either as they collect data or after collecting all data. Students will also need to draw free-standing tables, chairs, desks, and computers in the room. If the tables, chairs, desks, and computers are all the same, the students will turn these into "blocks" in AutoCAD. Students will need previous AutoCAD experience if this project is to be done via AutoCAD. If students are not experienced with AutoCAD or if two-dimensional design software is not available, students can use more graph paper to redesign the room. Students' main objectives in the library are to measure each wall, bookcase, desk, chair, doors, etc. and draw a bird's eye view and/or floor plan view of the area, both on scratch paper (if necessary for some students) and enlarged graph paper.

The teacher needs to remind students that they have to pick a scale. For example, one side of a graph paper square equals two feet. Some of the walls may come down to eighths and sixteenths of an inch. Such measurements can be difficult to represent on a graph paper, so the teacher may suggest drawing walls to the closest foot or half-inch but writing the correct dimensions down on the graph paper floor plan so they can be transferred to AutoCAD correctly.

The measuring of the library takes about seven 90-minute class periods. After all measurement has been completed and transferred to graph paper, students continue working in the classroom, where they finish up their floor plans on graph paper using their scratch papers and begin drawing the original floor plan in AutoCAD. Some groups may need to go back to the library to check their measurements. In this case, the teacher may prefer splitting the group in two. One group may go to the library to check the measurement while the rest of the group works on the AutoCAD floor plan. Transferring the graph paper floor plan to AutoCAD takes approximately three to four class periods, depending on the groups. After the original floor plan has been digitized in AutoCAD (and not only the walls, but also the free-standing tables, chairs, desks, and computers), each group member is assigned a section or plan for redesign. If there are a limited number of computers, students can work on the same computer and return one final AutoCAD design.

The teacher should facilitate rigorous-accountable talk and collaboration during each phase of the project. After the students have completed their redesign, they are to complete a PowerPoint presentation discussing their before and after floor plans and why they believe their redesign is conducive to learning. Group members should discuss the reasoning behind their redesign.

EXTENSION

As an extension to this project, students could calculate the area of the room and/or compare the library to historical libraries. Are they similar or different? Furthermore, students could calculate how many students it would take to fill up the floor space in the library if students had five feet between them. The students would turn these solutions in as a project, and they could use pictures to help define their solution strategies.

EVALUATION

Student presntations are evaluated using the Presentation Rubric contained in Appendix A and their designs are evalated using the Design Rubric in Appendix B.

EXAMPLES OF STUDENT WORK

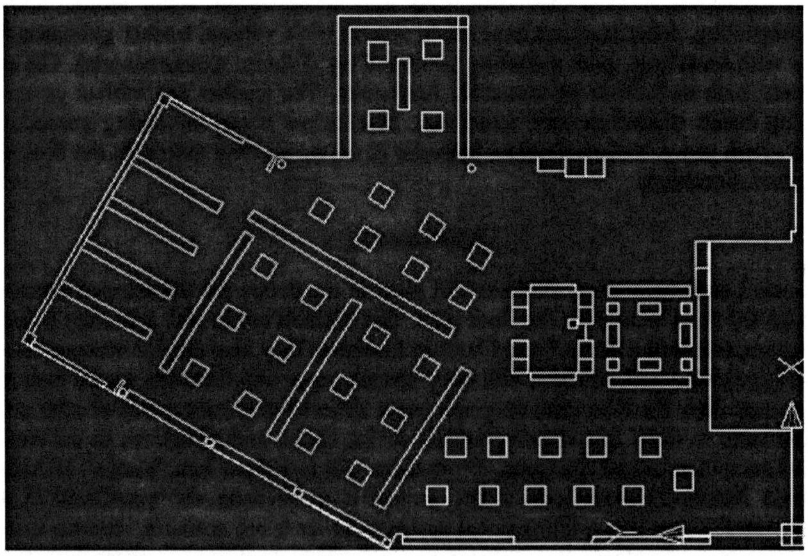

Figure 1. An example of student work: Library floor plan before students' redesign project.

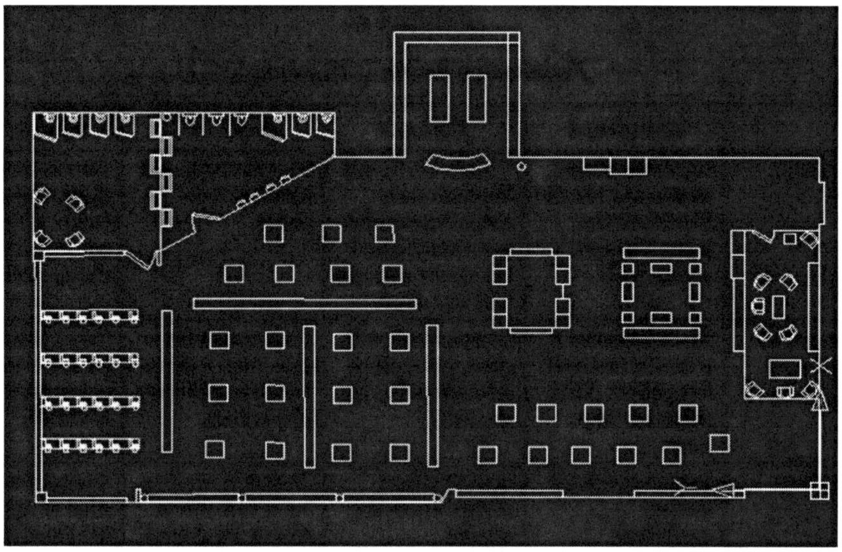

Figure 2. An example of student work: Library floor plan after students redesigned the library.

NOTES

[1] In our case, the school library has tall columns, and I wanted the students to use the string to find circumference in order to solve for diameter or radius when they started digitizing in AutoCAD. AutoCAD only uses diameter and radius, not circumference, for circles.

APPENDIX A

Presentation Rubric □ *16 points*

	Exemplary 4	*Mastery 3*	*Proficient 2*	*Introductory 1*
Use of Visuals	Highly developed, memorable visuals illustrate main point. Manages visuals well.	Visuals clarify and illustrate main point. Presentation is memorable and visuals are neat and well organized.	Visuals clarify and illustrate main points.	Visuals are not tightly linked to the presentation and do not support or clarify main points.
Communication of Main Points	Communication is articulate and well thought-out. Main points are very clear.	Communication is good, and most of the main points are covered.	Communication is somewhat lacking, but main points are still evident.	Communication is poor, and main points are not fully covered.
Body Posture	Student stands tall and has confidence.	Student almost always stands tall and has confidence.	Student is nervous, but is still able to convey the project.	Student is unsure what to say and is not proud of his/her work.
Fluency	Student does not stumble over words. The speech flows perfectly. Student does not use filler words.	Student does not stumble over words but uses a few filler words.	Student has a hard time getting the words out but eventually is able to communicate effectively.	Student consistently stumbles over words and uses filler words several times.

APPENDIX B

Design Rubric 16 points

	Exemplary 4	*Mastery 3*	*Proficient 2*	*Introductory 1*
All Project Materials Present	Group has graph paper, original design in AutoCAD, redesign in AutoCAD, and PowerPoint Presentation.	Group has most of the materials. Materials are somewhat disorganized, but present.	Materials are lacking and very disorganized.	Hardly any of the materials are present.
Original	Original drawing is dimensioned, complete, and truly a copy of the library.	Original has dimensions and is complete. There are only 1 or 2 mistakes.	Original design is present in AutoCAD but does not have dimensions. Several measuring mistakes are present.	Original does not have dimensions, is lacking, and has several errors.
Redesign	Redesign has dimensions and is original. Everyone in the group participated in the redesign. Redesign makes good use of space, and students used their imagination well.	Redesign has dimensions and is original. Most everyone in the group participated. Redesign makes good use of space.	Redesign is lacking but is present. Not everyone in the group participated. Redesign does not appear to differ much from the original.	Redesign is lacking. Not everyone in the group participated. Redesign does not make good use of space and is too close to the original.
PowerPoint Presentation	Graph paper, original AutoCAD design, and redesign are all included in the PowerPoint Presentation. Presentation clearly shows each group member's work.	All elements are present. Each group member's work appears in the presentation. Screenshots are minimal.	Elements are missing. Not all group members had a part in the PowerPoint presentation.	Elements are lacking. Some group members did not have anything to add to the PowerPoint presentation. There are no screenshots.

Jennifer Stitt
GIS Teacher,
Emmett J. Conrad High School, Dallas ISD

Serkan Özel
Educational Psychology,
Texas A&M University

LITERARY WORLD

MICHAEL "SKIP" GRAGERT AND KRISTIN SHAWN HUGGINS

19. THEY DID WHAT!?!?: GREEK GODS AND GODDESSES

WELL-DEFINED OUTCOME

Students will create a printed advertisement showing their mastery and understanding of 12 Greek gods' and goddesses' personalities, symbols, and spheres of influence as discussed in *The Odyssey* by Homer.

INTRODUCTION/BACKGROUND

Let's be honest. *The Odyssey*, for most students, is oh-so-boring. I want to pull my hair out (maybe that is why I shave my head) trying to keep the students engaged and interested. It isn't fair to the students. It isn't fair to the instructor. It REALLY isn't fair to the Greek gods and Homer. This project will catch the students' attention and "tenderize" their minds, allowing them to see the often untapped potential of studying ancient Greek culture. In short, the students will get extremely excited when they discover all the scandals and intrigues of the ancient Greek gods found in *The Odyssey*.

OBJECTIVES

Language Arts

- Produce a multimedia presentation with graphics and images that conveys a distinctive point of view and appeals to a specific audience.
- Determine, locate, and explore the full range of relevant sources, addressing a research question and systematically recording the information gathered.
- Clarify research questions and evaluate and synthesize collected information.
- Give presentations using informal, formal, and technical language effectively in order to meet the needs of the audience, purpose, and occasion; employ eye contact, speaking rate, volume, enunciation, purposeful gestures, and conventions of language to communicate ideas effectively.
- Participate productively in teams by building on the ideas of others, contributing relevant information, developing a plan for building consensus, and setting ground rules for making decisions.

World History

- Understand the history and relevance of major religious and philosophical traditions.
- Communicate in written, oral, and visual forms.
- Use problem-solving and decision-making skills and work independently and with others in a variety of settings.

Fine Arts

- Apply design skills to create practical applications, clarify presentations, and define choices made by consumers.

MATERIALS NEEDED

- Computer with Internet access and printer
- Microsoft Word (or an equivalent)
- Magic markers
- Scissors
- Plain printing paper or construction paper

R. M. Capraro, M. M. Capraro, J. Morgan and J. Scheurich, (eds.), A Companion to Interdisciplinary STEM Project-Based Learning: For Teachers by Teachers, 165–174.

- Greek gods and goddesses research sheet
- Greek gods and goddesses Formative Assessment I
- Grading rubrics

Safety Notes

Caution when using scissors.

Time frame for the project. The project should last four days.

DAY 1

Engagement

(15 minutes – meet in a computer lab with Internet access and an available printer)

In ancient Greece, people believed in more than twelve gods and goddesses . . . some acted very strangely by our standards. However, many students are unaware of the difference between monotheism and polytheism. A brief explanation of polytheism will pique the students' interest and also provide a base for students to proceed. Be prepared for many questions. Believe it or not, students may ask if the Greek gods are real. Without me realizing it when I created my lesson plans, this project challenged the belief system of some. So, I had to be very careful to point out that this is a religion from a LONG time ago. Students come with all levels of knowledge, so be prepared.

Exploration

(35 minutes)

Students will be given a research sheet (see Appendix A) containing 12 major Greek gods and goddesses, possible websites to research, and a few things for which to look. Although there are many more Greek gods and goddesses, having the students focus on these 12 will give students the opportunity to find background information helpful for *The Odyssey*. Furthermore, the students will be able to discover different aspects of ancient Greek culture based on these 12 deities.

Of course, this point in the project is excellent for laughing because the students will find out that Zeus married his sister and Apollo was bisexual. Get ready for "Ewwwww," "Yuck," and "Mister, why were they like THAT?" This section of the project is entitled "Exploration," and for the students, it is exactly that. They will explore and discover information about the ancient Greeks that they never knew prior to this opportunity. All of this will help shed some light on themes and topics in *The Odyssey*. Plus, it's just great fun to see the students' eyes light up when they discover how outrageously some of the ancients behaved.

The 12 main Greek gods and goddesses the students will explore are: Zeus, Apollo, Hestia, Hera, Ares, Poseidon, Hades, Aphrodite, Demeter, Dionysus, Hermes, and Athena.

Possible websites:	www.mythome.org/greek.html www.greek-gods.info/greek-gods www.mythweb.com/index.html
Research questions:	Of what was he/she the god or goddess? What were his/her symbols? Is there anything unusual about his/her family life? Who is his/her spouse? What was his/her personality like?

Homework. The students should continue their research, completing their research sheet before the next class period.

Note to the instructor. Students who are unfamiliar with Greek culture will ask questions about the historic context of the ancient Greek religion. In my experience with this project, several students have asked if the gods or goddesses are real and if their stories really happened. At first, it was very surprising to hear these questions, but the students were being genuine. A secondary benefit of this project is to see the students learn about how different other cultures can be from their own.

DAY 2

(15 minutes)

After the students have researched on their own, time should be given for students to teach students. Some students will have discovered things others have not. In groups of three or four, students should share their discoveries about the gods and goddesses with each other.

Evaluation

(10 minutes)

After the students teach each other, it is time for the first summative assessment (see Appendix B). This assessment will be a traditional paper and pencil exam requiring students to match the Greek gods with their symbols and domains.

Engagement

(25 minutes – meet in a computer lab with Internet access and an available printer)

Students will draw the name of one of the 12 main Greek gods or goddesses from a container as well as a product or service from a separate container. Finally, they will draw the name of his or her partner. The students will also receive a rubric and instructions for creating an advertisement for their product or service. Once the final project rubric (see Appendix C) has been discussed, the teamwork rubric (see Appendix D), which will be completed by the end of the third day, will be reviewed.

Some product or service examples:

- A watch
- A travel agency
- A fast food restaurant
- A used car
- Men's cologne
- Women's hairspray

The instructor can create a fun product list. It was great to see the football player in the room create an advertisement about pantyhose using the Greek god of the underworld, Hades. Don't be afraid to laugh and get ready to reassure each student that it is *okay* for a guy to create a lipstick advertisement.

The advertisement has the following requirements/design specifications:

- The advertisement should reflect the equal contributions of each partner.
- The owner of the company is the Greek god or goddess they chose, and a picture of him/her must be present.
- A symbol of the god or goddess must be shown.
- The advertisement must show the personality of the god or goddess. (For example, an advertisement containing Zeus would have a domineering tone to it, whereas Aphrodite's advertisement would have a lovely tone to it.)
- The advertisement should attempt to sell the product or service the way that god or goddess would attempt to sell it (i.e. forcefully, charismatically, or passive-aggressively).

Students will spend much of their time finding pictures and organizing their advertisement using Microsoft Word. When they finish their advertisement, they will print a copy, review it, and revise and edit it to improve their work.

Don't forget to remind students to save their work on either their network folder or a flash drive. Many students will not think to save their work unless they are reminded.

DAY 3

(50 minutes – entire class period)

On the third class day, students will continue working on their advertisements. During this day of the project, students get so engaged in their creations that they often do not pay attention to the time, so it might be helpful for the instructor to periodically remind them how much time they have left. Any work not finished on this day will be homework for the students.

DAY 4

Evaluation

On this day, students' advertisements will be evaluated based upon the requirements and design specifications. Students will present their advertisements to the class, explaining how they met each requirement in the rubric (see Appendix C).

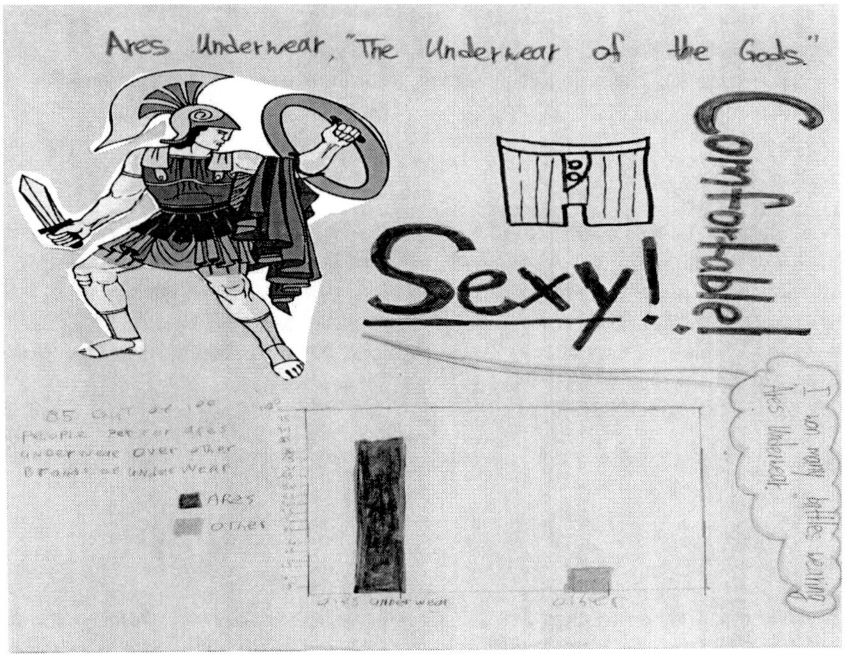

Figure 1. Although the above student-created advertisement has some organizational problems, it was one of the most entertaining. These students did a great job of showing Ares' personality as war-like and battle-focused. Additionally, this advertisement shows that the students clearly understood Ares' role in ancient Greek culture.

Extension

(Mathematics)

Getting the students' algebra or geometry teacher involved in the project would be a good way to create cross-curricular connections. One possible idea is to have the students create ratios. If they turned their 8.5 x 11 inch advertisement into a billboard alongside a road, how big would the pictures and lettering have to be in their advertisements?

APPENDIX A

Greek Gods and Goddesses Research Notes

Using the Internet and credible websites, research the following information about the gods and goddesses of Greek culture. Do not skip anything.

Zeus: domain: ______________________________

symbols: ______________________________

interesting family and/or spouse information: ______________________________

personality: ______________________________

Apollo: domain: ______________________________

symbols: ______________________________

interesting family and/or spouse information: ______________________________

personality: ______________________________

Hestia: domain: ______________________________

symbols: ______________________________

interesting family and/or spouse information: ______________________________

personality: ______________________________

Hera: domain: ______________________________

symbols: ______________________________

interesting family and/or spouse information: ______________________________

personality: ______________________________

Ares: domain: ______________________________

symbols: ______________________________

interesting family and/or spouse information: ______________________________

personality: ______________________________

Poseidon: domain:

____

symbols: ____

interesting family and/or spouse information: ______________________

personality:

____

Hades: domain:

____

symbols: ____

interesting family and/or spouse information: ______________________

personality:

____

Aphrodite: domain:

____

symbols: ____

interesting family and/or spouse information: ______________________

personality:

____

Demeter: domain:

____

symbols: ____

interesting family and/or spouse information: ______________________

personality:

____

Dionysus: domain:

____

symbols: ____

interesting family and/or spouse information: ______________________

personality:

____

Hermes: domain: ______________________________

symbols: ______________________________

interesting family and/or spouse information: ______________________________

personality: ______________________________

Athena: domain: ______________________________

symbols: ______________________________

interesting family and/or spouse information: ______________________________

personality: ______________________________

APPENDIX B

Greek Gods and Goddesses

Place the god or goddess on the line in front of the number next to his/her correct description. Next, write at least ONE of his/her symbols in the blank next to his/her correct description.

_________ 1. god of the sea and earthquakes; symbol ____________________

_________ 2. god of the underworld; symbol ____________________

_________ 3. messenger of the gods; symbol ____________________

_________ 4. god of weather; symbol ____________________

_________ 5. goddess of home and hearth; symbol ____________________

_________ 6. goddess of love and beauty; symbol ____________________

_________ 7. god of wine; symbol ____________________

_________ 8. goddess of the harvest; symbol ____________________

_________ 9. goddess of wisdom; symbol ____________________

_________ 10. god of war; symbol ____________________

_________ 11. god of light and the sun; symbol ____________________

_________ 12. goddess of marriage and birth; symbol ____________________

Aphrodite Demeter Ares Poseidon Hades Dionysus Hermes Athena Hera Zeus Apollo Hestia

Open-ended question: Examining the gods, goddesses, and their personalities, what conclusion can you draw about the daily lives of the ancient Greek people? Use evidence to support your conclusion.

____

____

____

____

APPENDIX C

Greek Gods and Goddesses Advertisement Rubric

	Exceeds Expectation 20 points	*Meets Expectation 15 points*	*Below Expectation 5 points*
Pictures of the Deity	The picture of the god or goddess clearly shows the deity. The deity is immediately recognizable without much difficulty.	The picture of the god or goddess is clear. The deity is recognizable, although it might take a moment or two to study the picture.	The picture is missing or is unclear. The audience cannot tell what deity is pictured.
Symbol of the Deity	An outstanding symbol of the deity is present. No guessing needed.	A symbol is present, although it might not be THE most common symbol.	The symbol is missing, unclear, or incorrect.
God's or Goddess' Personality	The ad clearly shows the correct personality characteristics. It is obvious what the deity was like without any guessing.	The ad shows the correct characteristics. It might take a couple of glances to understand the deity's personality.	The ad is vague or unclear as to what the deity's personality was like or the ad suggests an incorrect personality.
God's and Goddess' Sales Style	The ad excels in showing the deity's possible sales style. No guesses are needed, and the students are creative in getting the message across.	The ad correctly interprets what the deity's sales style might have been.	The ad is vague about how the god or goddess might sell the product or service.
Creativity	The advertisement is almost 100% original and is very appealing to see and examine.	The advertisement is more than 80% original and is appealing to see and examine.	The ad is less than 80% original and is not appealing to see or organized.

APPENDIX D

Gods and Goddesses Teamwork Rubric

	Exceeds Expectation 25 points	*Meets Expectation 18 points*	*Below Expectation 5 points*
Focus	No time was spent off-topic.	Most time was spent focusing on the project.	Considerable amount of time was spent off-topic.
Work Load	Both partners used each other's ideas to make the product better.	Both partners contributed to the success of the project.	One partner did more work than the other.
Teamwork	The team worked very well together. Any conflicts were dealt with professionally.	The team worked well together. When a conflict arose, the team members tried to work it out.	The team did not work well together. When a conflict arose, the team members argued.
Self-Paced	The team was able to use time efficiently.	The team was mostly able to use time efficiently.	The team had a hard time using time wisely.

Michael "Skip" Gragert
English Teacher,
Emmett J. Conrad High School, Dallas ISD

Kristin Shawn Huggins
Educational Administration and Human Resource Development,
Texas A&M University

MICHAEL "SKIP" GRAGERT AND KRISTIN SHAWN HUGGINS

20. THE PAPARAZZI: GETTING THE SCOOP ON THE SCHOOL

WELL-DEFINED OUTCOME

Students will collaborate with a classmate to conduct an interview and properly use quotes from that interview to create a well-written, informative article that is augmented by a visual representation.

INTRODUCTION/BACKGROUND

How can an instructor engage students with what they consider to be bland material? Some instructors ask themselves that question every year when it comes time to have students write in informative mode. Students often find informative writing dull, and their hearts are not in it. Also, many students have a hard time switching from informative writing to narrative writing to persuasive writing and so on. In a nutshell, students hate writing if they do not see the relevance.

The solution is to add "spice" to the assignment. Students love gossip (more precisely, students love outrageous gossip about others). To actively engage students, teach them informative mode writing, and encourage careful proofreading, the students write a tabloid gossip article over a teacher at their school. The topics and the project motivate students to put forth the effort needed to be successful.

OBJECTIVES

Language Arts

- Analyze, make inferences, and draw conclusions about informational texts. Appropriately use the informative mode to create an informational text.
- Edit drafts for grammar, mechanics, and spelling.
- Revise final drafts in response to feedback from peers and the teacher in order to publish written work for appropriate audiences.
- Participate productively in teams, building on the ideas of others, contributing relevant information, developing a plan for consensus-building, and setting ground rules for decision-making.

Media Literacy

- Analyze the role of media as sources of information, entertainment, persuasion, and education.
- Analyze strategies used by media to inform, persuade, entertain, and educate.
- Analyze techniques for producing media messages for specific purposes and effects.
- Develop skills for organizing, writing, and designing media messages for specific purposes and effects.
- Plan, organize, produce, and present media messages.

Journalism

- Plan and write relevant questions for an interview or in-depth research.
- Incorporate direct and indirect quotes and other research into copy.
- Demonstrate an understanding of the elements of news through writing.
- Gather information through interviews (in person or telephone).
- Write captions.
- Demonstrate an understanding of the function of headlines through the writing of headlines.
- Rewrite copy.
- Use illustrations or photographs that have been cropped to communicate and emphasize a topic.

R. M. Capraro, M. M. Capraro, J. Morgan and J. Scheurich, (eds.), A Companion to Interdisciplinary STEM Project-Based Learning: For Teachers by Teachers, 175–186.

Mathematics

- Select an appropriate representation (concrete, pictorial, verbal, or symbolic) in order to solve a problem.

Geography

- Use geographic tools to collect, analyze, and interpret data.
- Apply critical-thinking skills to organize and use information acquired from a variety of sources including electronic technology.
- Using problem-solving and decision-making skills, work independently and with others.

MATERIALS NEEDED

- Example tabloids for reference
- Computers with printing capabilities
- Digital camera for photographing teachers
- Instruction sheets
- Outline sheets
- Interview sheets
- Grading rubrics

Special considerations. Before the project can begin, the instructor must make sure other teachers are willing to participate and will not be offended if students write a gossip article about them. Of course, the instructor would ensure each topic is appropriate for school.

Time frame for the project. Completing this project will require five class days.

DAY 1

Engagement

Many students, especially English Language Learners (ELL), do not know what a tabloid is. In order to familiarize the students with tabloids, they will read two sensationalized articles from a tabloid newspaper. The entire class will discuss some of the headlines, topics, and quotes that were used in each article. Each student should make a few notes about how they could use the information discussed in an article of their own. The engagement will take 25 minutes.

Exploration

The instructor and students will review the tabloid project instruction sheet (see Appendix A) and rubric (see Appendix B). Then, the instructor will pair the students (in order to avoid conflicts and maximize each student's potential; see Appendix C). By pairing struggling writers with stronger writers, peer tutoring will take place on its own, leaving more time for the instructor to help those who need the most attention. The pairs will draw the name of a teacher (a teacher who has agreed to participate). Then, each pair of students will decide on the headlines for their articles. After the students have decided, the instructor should have some of the students write their ideas on the board; this will help the struggling students with ideas. The exploration will take 25 minutes.

Homework. Students should schedule an appointment with the teacher they drew for an interview to be conducted after the second day's class period. They should do this on the first day of the project to make sure the teacher will be available for the interview the second day. However, they want to make sure they interview the teacher after the second day's class period because important instruction and discussion concerning the interview process will take place during class. Therefore, it is important that students do not conduct the interview before this instruction and discussion has occurred.

DAY 2

The instructor and the students will review and discuss the article outline sheet (see Appendix D). Then, each pair of students will create a basic article plan for a rough draft. During this time, they will decide on topic sentences for each paragraph. After the outline is written, the groups will create four interview questions for their interviewee to answer on their project interview sheets (see Appendix E), keeping in mind their created headline and topic sentences.

Homework. Each pair must interview their teacher and record their answers on their project interview sheets before the next class meeting. Also, each pair should take at least one digital photograph of their interviewee, keeping in mind what pose and type of photo they will need for their final article (see Appendices B and F).

Special considerations. If there are limited resources regarding a digital camera, the instructor can take several photographs of each teacher participating and then have those students who cannot secure a camera choose one. Many of my students used their cell phones to take the photographs and brought excellent photos to use. For those students who excelled in this aspect of the project, it turned out to be a wonderful, laughable moment when they shared their "paparazzi" skills.

DAY 3

Explanation

Using the article outline sheets (see Appendix D) and the interview questions and responses (see Appendix E), the pairs of students will collaborate and write a rough draft of an article. The instructor should ensure that each pair of students is using at least two quotes from their interviewee (see Appendices A and B).

Several questions students should keep in mind as they write their rough drafts are:

- Who is my intended audience?
- What does my intended audience already know about the topic and/or teacher?
- What vocabulary do I need to use to make sure my article sounds professional and believable?

After their rough drafts are complete, the students should reread their writing and correct any mistakes. During the revising and editing step, the students will review the rubric (see Appendix B) to make sure that they have met all the requirements found on both the rubric and the project information sheet (see Appendix A) as well as group with another pair of writers to trade rough drafts and gain peer editing suggestions.

This is a very important step. Many students will rush through the revising and editing phase. It is vital that the instructor slow them down. Perhaps the instructor and the pair revise and edit the first draft together. Also, many of my students did not know what to look for when revising and editing. It might be helpful to put a quick list of things editors might correct on the board in the front of the classroom. For example, a simple list would include: capitalization, spelling, punctuation, word choice, etc. This part of the project will take an entire 50-minute class period.

DAY 4

In pairs, the students will use their revised rough drafts to type their final drafts. The pairs will place the digital photograph they took of their interviewee in their article to maximize effect, based on samples of tabloid articles they have read. It might be helpful to point out that their article should look like an article in a tabloid, not an essay with a picture at the beginning or end. Some students may be unsure how to insert pictures in a word processing program. So, the instructor and other knowledgeable students should assist when necessary.

Once the final draft is typed, the students should reread their work and revise and edit one last time to ensure that they have eliminated all errors. To help with this task, they should reread their peer editors' suggestions from the previous day. Also, they should double check to make sure they have met all of the requirements found in the rubric.

Once they have double-checked their work, they should save their work and print two hard copies, one for the instructor and one for themselves. It is a good idea to have them save their work, either on their network profiles and/or to a flash drive. Many students will not think to save their work unless they are reminded. This part of the explanation will take another 50-minute class period, and the class will need to meet at a location where students will have access to a computer and printer.

DAY 5

Evaluation

The students, still in their pairs, will read another group's article in class and will give it a grade based on the rubric. After each group has finished grading, each group will present the article they read to the entire class. Then, they will convey why they gave it the grade that they did. After each pair presents the article it graded, a short debate might be necessary if other students feel that the article deserves a different grade. This is where instructor mediation is important in order to allow an honest, yet productive, discussion about the article's quality.

There are three different, effective ways to incorporate the students' grading:

75% of the grade from the instructor, 25% of the grade from the students. This form of grading allows students to have some input into the grades without overriding the instructor. This form might be useful if students are new to the English language and struggle with the "basics." Also, this way is best if you have a class of students who are often "rowdy" or unfair.

50% of the grade from the instructor, 50% of the grade from the students. This form allows the students to have a considerable say in the grades, which places trust in their judgment. This is definitely not for an overly critical or unfair class; however, with the right class it will create a 50/50 relationship between the instructor and the students.

25% of the grade from the instructor, 75% of the grade from the students. This form puts the burden of grading on the students. They are required to take responsibility for fair and impartial grading. This form places much trust in the students' judgment.

Any one of these three forms places significant responsibility on the students; no longer can they be passive observers, but they must take an active role in their peers' and their own performance in class. By having students' input on the grading, it sends a message to the class that education is a community event, not just an instructor event. It encourages each student to take ownership of his or her education and evaluation. It also eliminates some of "the teacher wasn't fair to me" excuses that some students might attempt to use. The evaluation by the students will use the rubrics in the appendices (B through F), and will take an entire 50-minute class period.

"Local High School Teacher Raised By Platypuses!"

Ms. [redacted] has a secret wet life. "Have you notice she walks on Her hands and feet? She only Eats raw fish and toe jam. Ms. [redacted] skin color is a Green and brownish color. It helps her camouflage from Her natural predator the goldfish and earthworm.

On her
asked her some
told her to
color of her
have a duck
that's way my
big".

interview we
questions. W
explain her
skin. "I might
back ground
lips are so

We also asked her how she eats with her beak. "I really haven't recognize that because of my over size floppy lips." Ms. [redacted] lips are so huge to get to places she has to have a helicopter lift her by her lips.

Figure 1. The above is an example of a completely inappropriate, failing article. The students in the class recognized right away the unacceptable length and insulting content of the article. The students who worked on this article put little to no effort into creating a quality product. They certainly did not comply with the instructions or final project rubric, and the other students in the class graded it so. In this case, it was extremely helpful to have the students grade each other's work, for peer pressure is stronger than instructor pressure. The students who wrote this article ended up apologizing to the teacher for being so mean-spirited. They also voluntarily rewrote the article.

Figures 1–3 are examples of articles written by pairs of students; in each of the pairs, one of the students was a decent writer and the other was a struggling ELL writer. Although there were a few unacceptable articles, the vast majority were great accomplishments. For nearly all of the ELL writers, this project allowed them to create the best piece of writing they have ever composed.

THE SIX...HUNDRED DOLLAR WOMAN

Article by ████ and ████ March 12, 2009

On February 13, 2009, Ms. ████ was jogging like she does daily, when a drunk driver spinned out of control and pinned her into a tree, where she spent countless hours. Nobody was around to help. However, ████ was found by an evil man. "who?", you say. It's Dr. Doom. Dr. Doom found her nearly dead.

Following the accident, Dr. Doom rushed her to his lab and put her under the knife. The rebuiltment took hours. "I was in a lot of pain, couldn't walk, and took many pain pills," ████ says. As she recovered quickly, she learned to use her new powers.

Then, Dr. Doom sent ████ to the Pentagon to destroy the U.S. military's secret weapon. Although she didn't find it, she destroyed most of the Pentagon. Witness's caught the devastation on video. One reporter at the scene says, "I've... never...seen...anything like this not since Dr.Doom disappearance."

████ terrorizing the Pentagon.

The U.S. needed help so the Fantastic Four were alerted by President Obama. By the next time ████ struck, the Fantastic Four had already defeated her and put her in custody of the Supreme Court. She was asked, "What made you betray the U.S.?", and responded, "I did it, so my boss (Dr. Doom) would become the most powerful people on Earth. " Of course, stronger words from a strong woman.

Right before ████ was put away, we asked her, "Any words for the Fantastic For? " and she responded, "They'll feel my wrath." The Fantastic Four shouldn't worry... for now.

Dr. Doom, while watching ████ being defeated by the Fantastic Four.

Figure 2. This is an example of an acceptable article. There are several mistakes with the conventions (punctuation, grammar, etc.). However, the mistakes do not interfere with the "readability." The other students in class gave this a "B" grade. One thing they pointed out was that there was not a picture of the actual teacher, just some photographs the students found on the Internet. Although it is entertaining, the article is missing some information. But overall, this is a very good beginning to a potentially excellent article.

SHOCKING NEWS ASSISTANT PRINCIPAL RAISING AN ARMY...OF ZOMBIES!!!

█████, █████

ASSISTANT PRINCIPAL SEEN IN THE CONFERENCE ROOM WITH ZOMBIES!!!

Assistant principal, Mrs. █████ has been seen in the █████ High school conference room with what appears to be the leader of a zombie army! Neighbors say that she has been having meetings at her house at midnight for the last three months. Her co-workers say that she has been acting very different.

"No one can stop me," █████ shouted to a group of ill-tempered youth. "You shall all be my minions to conquer Dallas, then Texas, then THE WORLD!!!"

WHAT IS MRS. █████ HIDING FROM US?

Last week we tried to interview her but she kept getting away from us. We have visited her house to ask her a few questions but her roommate says that she no longer lives there. What is Mrs. █████ hiding from us?!

MRS. █████ HAS BEEN ARRESTED FOR TRYING TO TAKE OVER THE WORLD WITH AN ARMY OF ZOMBIES!!!

On March 5, 2009 around one 'clock in the morning Mrs. █████ former assistant principal was convicted of trying to take over the world with an army of zombies. A S.W.A.T. officer dressed as a zombie and gave his captain the information that they needed so they can arrest her. She went to try on the 11th of March and was sentenced to LIFE in prison!!! Not only did she try to take over the world but, she also killed 11 teenage students and turned them into zombies!!!

Figure 3. This article was graded as going above expectations. The students who graded it pointed out how it actually looks like an article. Additionally, they pointed out that although there was only one picture (which was not of the assistant principal), the article had very few problems with conventions. Additionally, the article is appealing in entertainment and aesthetic value and shows the students spent time revising and editing.

EXTENSION

After the students have turned in their own informative news articles, the student will attempt to find a tabloid article that has already been published that deals with a topic similar to their article. After finding the article, the students will locate the thesis statement and topic sentences of the article as well as grade the article using the final project rubric (see Appendix B).

APPENDIX A

Tabloid Instruction Sheet

Tabloid: A gossip magazine; a small newspaper with sensational material.

GOAL:

You and a partner will conduct an interview and write an error-free article to be used in a class tabloid.
YOU'LL BE PUBLISHED!

Mode:

Informational

Grading:

20% of your six weeks grade:
- 5% for your interview and rough draft (see rubric)
- 5% for partner's participation (see rubric)
- 10% for your final draft (see rubric)

Requirements:

*Rough draft (written) and final draft (typed); both must be between 200–300 words.
*Must use at least 2 quotes from your interviewee.
*Must be entertaining.
*Must have at least ONE picture.

Steps:

1. Choose your topic and interviewee.
2. Create an outline.
3. Create four interview questions.
4. Interview the person.
5. Write your rough draft.
6. Revise and edit.
7. Create your final draft (must be typed).
8. Turn it in like this: Final Draft and rubrics as one packet; Outline, interview sheet, and rough draft as a separate packet.

Dates:

NO LATE WORK ACCEPTED!!!!!

Date:	Topic, 4 questions, outline
Date:	Interview sheets due at the beginning of class
	Rough draft due at the end of class
	Revising and editing due at the end of class
Date:	Final draft, ready to publish and free of errors, due at the end of class

APPENDIX B

Final Draft Rubric

Inappropriate topics, terms, or demeaning language will result in a failing grade.

Name: ______________________________

Name: ______________________________

Headline: ______________________________

	Insufficient *0 points*	*Almost There* *10 points*	*Sufficient* *20 points*	*Excellent* *25 points*
Informative Mode	More than 40% of the article was not in informative mode.	More than 25% of the article was not in informative mode.	Less than 25% of the article was not in informative mode.	Less than 5% of the article was not in informative mode.
Article Content	Five or more sentences were not professional. The article was less than 200 words.	Four or fewer sentences were unprofessional. The article was less than 200 words.	The article was almost entirely professionally written. The article was between 200 and 300 words.	The article was entirely professionally written. The article was between 200 and 300 words.
Revising and Editing	7 or more grammar, spelling, or punctuation mistakes.	4–6 grammar, spelling, or punctuation mistakes.	2–3 grammar, spelling, or punctuation mistakes.	1 grammar, spelling, or punctuation mistake.
Pictures	The pictures were irrelevant to the article, or none were present.	The pictures were only somewhat related to the article.	The pictures were related to the article and provided a good visual aid.	The pictures were related to the article, and they clearly supported the article's message.
Quotes	No quotes from the interview were included.	One or two quotes were present. The quotes were irrelevant to the article.	At least two quotes were present. The quotes were relevant to the article.	At least two quotes were present. The quotes were excellent in making the article seem professional.

APPENDIX C

Partners' Work Rubric

Name: ____

Name: ____

Headline: ____

	Insufficient 0 points	*Almost There 10 points*	*Sufficient 20 points*	*Excellent 25 points*
Listening	Neither partner worked well together at all.	The partners attempted to listen but did not seriously consider each other's ideas.	The partners did a good job listening to each other and using each other's ideas.	The partners listened to each other and played off of each other's ideas. There was true collaboration.
Time Management	Most of the time was spent off-topic.	Some of the time was spent off-topic.	Very little time was spent off-topic.	Almost none of the time was spent off-topic.
Teamwork	One person did most of the work on the article.	One partner contributed more than the other partner.	Both partners contributed to the article.	The partners both contributed equally to the success of the article.

APPENDIX D

Article Outline

Name: ____

Name: ____

Headline: Make it catchy, fun, and 5–10 words.

HEADLINE: ____

Paragraph #1: Introduction: At least 4 sentences; the last sentence is your thesis statement.

THESIS: ____

____

Paragraph #2: First Body Paragraph: At least 5 sentences; the first sentence is a topic sentence.

TOPIC SENTENCE: ____

____

Paragraph #3: Second Body Paragraph: At least 5 sentences; the first sentence is a topic sentence.

TOPIC SENTENCE: ____

____

Paragraph #4: Third Body Paragraph: At least 5 sentences; the first sentence is a topic sentence.

TOPIC SENTENCE: ____

____

Paragraph #5: Conclusion: At least 3 sentences.

____

____

____

____

APPENDIX E

Interview Sheet

Name: ______________________________

Name:

____

Person to interview: ______________________________

Headline:

____

Question #1: ______________________________

His/her response: ______________________________

____

Question #2: ______________________________

His/her response: ______________________________

____

Question #3: ______________________________

His/her response: ______________________________

____

Question #4: ______________________________

His/her response: ______________________________

____

APPENDIX F

Interview and Rough Draft Rubric

Inappropriate topics, terms, or demeaning language will result in a failing grade.

Name: ____________________

Name: ____________________

Headline: ____________________

	Insufficient *0 points*	*Almost There* *10 points*	*Sufficient* *20 points*	*Excellent* *25 points*
Interview (A Teacher)	The interview was not completed.	The interview was completed by only one partner.	X	Both partners conducted a complete interview.
Interview Questions	The questions had little or nothing to do with the article.	Only half the questions were relevant to the article.	Only one question was irrelevant in relationship to the article.	Every question was designed to add meaning to the article.
Rough Draft	The rough draft was less than half complete.	The rough draft was incomplete.	The rough draft was nearly complete.	The rough draft was entirely complete.

Michael "Skip" Gragert
English Teacher,
Emmett J. Conrad High School, Dallas ISD

Kristin Shawn Huggins
Educational Administration and Human Resource Development,
Texas A&M University

MICHAEL "SKIP" GRAGERT AND KRISTIN SHAWN HUGGINS

21. DON'T USE THAT TONE WITH ME!: HOW TO INTRODUCE A DIFFICULT LITERARY CONCEPT

WELL-DEFINED OUTCOME

Students will write an error-free, original letter through the writing process, using a particular tone, knowledge of the target audience, and appropriate rhetorical devices necessary to persuade. The students will send the letters to researched, targeted parties.

INTRODUCTION/BACKGROUND

Moving freshmen English students from the concrete to the abstract can be challenging at best. Oftentimes, freshmen have a hard time correctly identifying the tone of a piece of literature, and they have an even more difficult time creating tone with written language.

One solution is to assign a relevant and interesting project to the students. If a student has a personal stake in the project, she/he will give much more effort. To engage the students, teach them how to recognize tone, and give them the opportunity to create different tones in their own writing, the students will write a professional business letter in a specific tone to a company of their choice with the intent of getting a reply.

OBJECTIVES

Language Arts

- Evaluate changes in formality and tone within the same medium for specific audiences and purposes.
- Write persuasive texts to influence the attitude or actions of a specific audience on specific issues.
- Structure ideas in a sustained and persuasive way and develop drafts that include transitions and rhetorical devices to convey meaning in open-ended situations.
- Understand the function of and use the conventions of academic language in writing.
- Use appropriate capitalization and punctuation conventions in compositions.
- Spell correctly; use various resources to determine and check correct spelling.

Business Communications

- Organize ideas logically and sequentially.
- Locate and interpret written information.
- Compose persuasive messages.
- Compose business letters using appropriate organizational strategies.

Geography

- Use geographic tools to collect, analyze, and interpret data.
- Apply critical-thinking skills to organize and use information acquired from a variety of sources including electronic technology.
- Using problem-solving and decision-making skills, work independently and with others.

MATERIALS NEEDED

- The book *Idiot Letters* by Paul Rosa (ISBN 038547508X)
- Computer with Internet access and printing capabilities

R. M. Capraro, M. M. Capraro, J. Morgan and J. Scheurich, (eds.), A Companion to Interdisciplinary STEM Project-Based Learning: For Teachers by Teachers, 187–195.

- An envelope
- Correct postage
- Tone words
- Paper
- Grading rubrics

Time Frame for Project. The teacher will need to allow four days for this project.

DAY 1

Engagement

Using students' prior knowledge to introduce a new concept is usually a good idea. It brings all students into discussion because their own personal experiences make it easy for them to talk, discuss, and learn from each other. For this exercise, a good list of tone words can be found at *http://dothgrin.net/* tonewords.pdf.

First, the instructor should write a light-hearted tone word on the board, such as *whimsical*, *amused*, or *fanciful*. Many students will look at the words, scratch their heads, and ask, "What do *those* words mean?" Perhaps a student will know what they mean and can explain them to the class, or if not, the instructor will. Ask the class to share a quick story about a time they felt whimsical or fanciful. After each student has shared, write down a list of three to five words or phrases the students used in their stories that showed that tone.

Amused	Whimsical
Told a joke	I was daydreaming
Smiling	...could never happen
Laughed	Wished

Point out that each of these phrases contributes to the "feeling" of that particular tone.

Next, do the same exercise with negative tone words, such as *inflammatory*, *obnoxious*, or *arrogant*. This part of the project should take 15 minutes.

Exploration

The book *Idiot Letters* by Paul Rosa is a great way to see different tones and analyze writing. The author writes outrageous letters to companies in order to get a response. The instructor chose two of Rosa's letters and their responses to show together to provide examples of distinct differences in tone. Any one of the letters would be great to use; however, these two examples were selected because students can easily identify the tones of each letter.

The first example is Rosa's letter to CITGO (pp. 21 & 22) regarding an advertisement and the company's response; the letter is *exaggerated* and *sarcastic* (in a humorous way). CITGO plays along in their response, and they even congratulate Rosa for making them laugh.

The second suggestion is the letter written to Merriam-Webster (pp. 67 & 68). Again, the tone is *sarcastic* and *facetious* (humorous, not condescending), and the response Rosa receives is *dry*, *serious*, and *straightforward*. The students' review of these two letters and their responses will let them see not only how different tones are created using rhetorical devices such as sarcasm, exaggeration, and ridiculous situations, but also short, to-the-point sentences in straightforward responses. The students will laugh, and when they read the letters and responses, they will say collectively, "Ahhh! They didn't pick up on the author's tone." This exploration should take 30 minutes.

Homework. The students should choose a company to whom they want to write a letter, find the mailing address of the company, and bring an envelope and stamp to the next class period. This might seem like a simple task; however, if the students do not find the mailing address on their own, they will spend too much valuable time on Day 2 looking up addresses on the Internet.

Special considerations. Do not overestimate how many students actually know how to properly address an envelope. With text messaging and the Internet, envelope addressing has become a forgotten skill.

DAY 2

Explanation

Prewriting. Upon entering class, students will receive a tone word. This is the tone in which they must compose a letter to the company they have selected. The tone words should be simple enough for the students to understand and emulate.

The following is a sample list of tone words to use:

Cheerful	Friendly	Playful	Optimistic
Aggravated	Desperate	Irritated	Outraged
Baffled	Inquisitive	Pleading	Urgent

On a sheet of paper, students should brainstorm how they could incorporate their tone in a professional business letter. They should define their tone word, identify synonyms, and perhaps free write for a few minutes in order to begin generating some ideas.

Special considerations. It might be beneficial for the teacher to model what free writing is in order for the students to see how to freely generate ideas on paper.

First draft. After the students have formed some thoughts through brainstorming, they can begin their first draft. Before they begin their draft, though, the instructor should hand out the grading rubric (See Appendix A and B) and go over expectations for the final letter. After the rubric explanation, the students should go to http://owl.english.purdue.edu/owl/resource/653/01/ and use the information there to begin their first draft. The website gives great hints and instructions on what to say and how to organize a letter. Reviewing *Idiot Letters* for stylistic analysis is also a good idea.

At the end of the class period, students should save their work either on a flashdrive or on their network folder after they have printed a copy for themselves. Prewriting and writing a first draft will take an entire 50-minute class period.

Homework. Students should revise and edit their letters, eliminating any errors.

DAY 3

Evaluation

Peer review. As students enter, have them retype their revised copy of their letter, print an updated version, and have another student in the class review their work. At this point, the students should write some useful and helpful feedback on the letters based upon the final grading rubric. The rubric grades them on the peer review as well as their own edit and revision of their work (see Appendix C).

Here are examples of useful and not useful feedback:

- Not useful feedback: "Your letter is good."
- Useful feedback: "I like how you use humor in the first sentence to grab the reader's attention."
- Not useful feedback: "You have a lot of mistakes."
- Useful feedback: "The three comma splices in the last paragraph make it hard to read. Here are my suggestions..."

Special considerations. The instructor may need to give a thorough explanation and possibly model peer reviewing for the students, especially if this project is the first time the students have utilized peer reviews. The reviewing process will take an entire 50-minute class period, and the students will need access to computers with the Internet and a printer.

DAY 4

Evaluation

Final draft. After a peer has reviewed their letter, students should seriously consider suggestions that will improve the quality of their letters. At this point, students should type their final draft. Upon completion, students should save their work to a flashdrive or to their network folder and also print two copies, one for the instructor and one for themselves.

After they have signed their letter, they should address their envelope, carefully fold and place their letter in the envelope, and stamp it. Students can mail their own letters, or the instructor can mail the entire class' letters for them to ensure they are actually mailed. (See evaluation rubrics in Appendices A, B, and C).

Dallas, TX

Burger King Corporation
5505 Blue Lagoon Drive
Miami, Florida 33126

Februrary 19, 2009

Dear Sir or Madam,

I am a simple man. I work and go to school to better myself and help pay the bills in my house. I love eating at Burger King. I love taking my little brother and sister to Burger King to buy a Whopper. My family is also about to go bankrupt, and we need your help.

Have you ever heard a child cry themselves to sleep because they were not able to eat dinner? Have you ever had to tell a child, "no, we cannot go to Burger King because we have no money?" Well, I have. I have to tell my little siblings that must go without the nourishing double meat, double cheese Whoppers that sustain their life functions for another day.

The help I need is assistance, just in the short term, to help combat this horrible recession that we are in. With gas prices, food prices, and housing costs rise, I am unable to afford a double meat, triple meat, or ANY meat Whopper. If you could help keep my little sister and brother able to eat the Whoppers, it would be greatly appreciated.

Sincerely,

Figure 1. The above is a letter written to Burger King. The student who wrote this letter did a wonderful job of creating emotional appeal, using sympathy to try to convince the company to help pay for Whoppers. He understood how to incorporate a desperate or pleading tone.

ne
Dallas, TX

Sea World Orlando
7007 Sea World Dr.
Orlando, FL 32821

February 19, 2009

Dear Sir or Madam:

I visit your amusement park every year to get away from this cruel life and finally be able to relax. When I visit this wonderful place I buy souvenirs. Also I go to see Shamu. Sometimes this visit starts to become very expensive and personal problems occur.

I buy your souvenirs and take pictures for precious memories, but I think the prices should be lowered, especially for the food. If your prices were lowered, then it would be more economical and the financial situations would disappear (especially a person who suffers from extreme credit problems like me). Every time I go to see Shamu perform, the water that he splashes ruins my new hair extensions (better known as a weave).

I am not the only one who suffers from these problems. Do you think that you can make this possible? I would be very thrilled if this happens. I promise that you would have more than 200 members of my family visiting you site. I look forward to your extravagant reply.

Sincerely,

P.S. Please send me a hat (so I can protect my hair) and a souvenir key chain.

Figure 2. The above is an example of a letter written to Sea World. Although the letter has a few errors, the student was able to be humorous and exaggerated in parts. However, this student did not really understand how to incorporate tone in the letter. Figure 3 contains the response the student received.

Extension

It is good fun to see the tone of each of the response letters the students receive. This is a great way to continue examining tone after the project is over. As the letters come in, photocopy each one and have the class identify the tone of the response letter to identify whether or not the company chose a serious or light-hearted tone. Also, it is fun to see if the company picked up on the tone the students used in their original letters. The extension takes place as the responses to the students' letters arrive.

SeaWorld
ADVENTURE PARK
Orlando

7007 SeaWorld Drive · Orlando, Florida 32821

March 25, 2009

Ms. [redacted]
Dallas, TX [redacted]

Dear Ms. [redacted]

Thank you for taking the time to share the experiences you encountered on your visits to SeaWorld Orlando. I always appreciate hearing from our guests and the opportunity to respond.

We understand the tough economic times are affecting many families. We do our best to provide the best value for price and feel we are comparable to other theme parks in the Orlando area. We continually evaluate our products and pricing to maintain a competitive advantage.

At Shamu Stadium, we do have splash and non-splash zones. Our guests get to choose their seats in one area or another, depending upon whether they want to be wet or not. I would suggest sitting in one of our drier areas on your next visit.

Thank you again for sharing your comments and concerns. We look forward to your future visits.

Sincerely,

Rusty Harman
Senior Director
Customer Relations

Figure 3. The above is the response to the student letter in Figure 2.

APPENDIX A

Organization, Content, and Conventions Final Letter Rubric

	Not nearly enough	*Almost enough*	*Sufficient*	*Superb*
Parts of the Letter 15 points	The letter is missing three or more key elements such as the date, salutation, or inside address. 0 points	The letter is missing two key elements such as the date, salutation, or inside address. 5 points	The letter is missing only one key element such as the date, salutation, or inside address. 10 points	The letter has all of the key elements including the date, salutation, and inside address. 15 points
Organization 20 points	The letter is not organized in any logical fashion. Even after reading it several times, it is unclear. 0 points	The letter is somewhat organized; however, the reader has to read it several times to understand it. 10 points	The letter is organized enough to make sense; however, it does not have enough transitional elements to make it flow. 15 points	The letter has no organizational problems. Each paragraph flows into the next. 20 points
The Main Point 25 points	The main point is unclear or is missing. 0 points	The main point is present; however, there are no details to justify it. 12 points	The main point is clear and justification for the issue or concern is provided. 20 points	The main point is well-stated; multiple ideas are present to justify the issue or concern. 25 points
Professionalism 25 points	The letter is not in block style, and the word choice is not professional. 0 points	The letter is either not in block style, or the word choice is not professional. 12 points	The letter is in block style, and the word choice is at least 80% professional. 20 points	The letter is in block style, and the word choice is 100% professional. 25 points
Conventions 15 points	The letter has more than six mistakes in grammar, punctuation, or spelling. 0 points	The letter has four to five mistakes in grammar, punctuation, or spelling. 5 points	The letter has two to three mistakes in grammar, punctuation, or spelling. 10 points	The letter has one or no mistakes in grammar, punctuation, or spelling. 15 points

APPENDIX B

Tone, Rhetorical Devices, and Target Audience Final Letter Rubric

	Not nearly enough	*Almost enough*	*Sufficient*	*Superb*
Identifying Tone 25 points	The student does not demonstrate that he/she can properly identify the tone of one other student letter. 0 points	The student identifies the tone of one other student letter. 12 points	The student identifies the tone of three other student letters. 20 points	The student identifies the tone of five or more of the other students' letters. 25 points
Using Tone 25 points	The tone of the student's letter is not recognizable even after reading it twice. 0 points	The tone of the student's letter is recognizable only after reading it a second time. 12 points	The tone of the student's letter is clear because of one or two well-written sentences. 20 points	The tone of the student's letter is clear because of three or more well-written sentences. 25 points
Persuasive Rhetorical Devices 25 points	The student's letter does not contain any sentences that attempt to persuade. 0 points	The student's letter has one sentence that attempts to persuade. 12 points	The student's letter has two sentences attempting to persuade. 20 points	The student's letter has three or more sentences attempting to persuade. 25 points
Target Audience 25 points	The letter's content does not match the student's target audience. 0 points	The letter has one or two sentences that match the target audience. 12 points	The letter has three to five sentences that match the target audience. 20 points	Every sentence matches the target audience. 25 points

APPENDIX C

Peer Review and revising and Editing Rubric

	Not nearly enough	*Almost enough*	*Sufficient*	*Superb*
Focus 25 points	Student focused less than 50% of the time on seriously reading and critiquing. 0 points	Student focused 50% of the time on seriously reading and critiquing. 12 points	Student focused 75% of the time on seriously reading and critiquing. 20 points	Student focused all of the time on seriously reading and critiquing. 25 points
Helpful Comments 25 points	The student's comments were unhelpful and vague. 0 points	Only one of the comments was helpful to the author in revising. 12 points	Two to four comments were helpful to the author in revising. 20 points	Five or more comments were helpful to the author in revising. 25 points
Revising 25 points	The student did not revise the rough draft at all. 0 points	The student made very little revision to the rough draft. 12 points	The student made revisions to the rough draft but could have done more. 20 points	The student made thorough revisions to the rough draft. 25 points
Editing 25 points	The student did not seriously edit the rough draft. 0 points	The student attempted to edit the rough draft but did not finish. 12 points	The student edited the rough draft but missed a few errors. 20 points	The student thoroughly edited the rough draft, eliminating every error. 25 points

Michael "Skip" Gragert
English Teacher,
Emmett J. Conrad High School, Dallas ISD

Kristin Shawn Huggins
Educational Administration and Human Resource Development,
Texas A&M University

FOOD SCIENCE

SENCER CORLU AND KATHY VONDRACEK

22. CHEMISTRY OF BAKING

WELL_DEFINED OUTCOME

The student will conduct a comprehensive research and create an effective essay to explain the chemistry of cooking.

INTRODUCTION

All baking is based around the use of wheat flour, but many other ingredients are also used, each of which have different effects on the final product. This project-based learning (PBL) activity will help students discover the chemistry of baking and learn about physical and chemical change, balancing a chemical equation, ideal gas laws, and acid-base reactions. Students will also have the opportunity to talk about quality, taste, and hygiene issues surrounding the food we eat. They will also practice converting units (dimensional analysis) and manipulate direct and indirect proportions.

OBJECTIVES

Scientific Process

- The student should be able to analyze, review, and critique scientific explanations, including hypotheses and theories, as to their strengths and weaknesses using scientific evidence and information. Science is a way of learning about the natural world.
- Students should know that science has built a vast body of changing and increasing knowledge described by physical, mathematical, and conceptual models and also that science may not answer all questions.
- The student will use scientific methods during field and laboratory investigations.
- The student is expected to plan and implement investigative procedures including asking questions, formulating testable hypotheses, and selecting equipment and technology and make responsible choices in selecting everyday products and services using scientific information.

Chemistry

- Students should be able to conduct field and laboratory investigations, use scientific methods during investigations, and make informed decisions using critical thinking and scientific problem solving.
- Students will study a variety of topics that include: characteristics of matter, energy transformations during physical and chemical changes, atomic structure, the periodic table of elements, behavior of gases, bonding, nuclear fusion and nuclear fission, oxidation-reduction reactions, chemical equations, solutes, properties of solutions, acids and bases, and chemical reactions. Does this unit really do all that?
- Students will investigate how chemistry is an integral part of daily life.

Personal and Family Development

- The students will determine career options related to nutrition, food science, and wellness and demonstrate basic meal-management techniques.
- The students are expected to identify the impact of technology on meal management, demonstrate basic principles of sanitation and safety relating to meal management, and apply management techniques when planning and preparing simple meals and recipes.

R. M. Capraro, M. M. Capraro, J. Morgan and J. Scheurich, (eds.), A Companion to Interdisciplinary STEM Project-Based Learning: For Teachers by Teachers, 199–205.

Algebra

- Students should be able to work with equations and inequalities as a way of asking and answering questions involving functional relationships.
- Students should be able to set up equations and inequalities and use a variety of methods to solve them.

Language Arts

- The student will be required to organize ideas in writing to ensure coherence, logical progression, and support for ideas. The student will use prewriting strategies to generate ideas, develop voice, and plan. Finally, the student will compile written ideas and representations into reports, summaries, or other formats and draw conclusions.

MATERIALS NEEDED

Experiment 1. A pipette, Bunsen burner, triple beam balance, baking soda, vinegar, test tubes, and enough rubber balloons.

Extension Experiment. Two rubber balloons that are the same size but different colors (blue and red), a tape measure, two beakers, hot water, ice, and a thermometer.

Safety Procedures

Wear goggles and aprons, tie hair back, and wash hands (before, during, and after).

STUDENT INTRODUCTION

After a hard day at school, you arrive home and your mouth begins to water when you encounter the delicious smell of your mom's home cooking, whether cakes, cookies or some other baked snack. This irresistible smell and the delicious baked goods are the results of the interaction of various chemical and physical processes, which are controlled by your mom. Unless your mom is a chemist, it is very likely she's unaware of the science and mathematics behind her cooking skills. As you eat your cookies, you hear your mom complaining about your little sister, one of a curious mind, who is annoying her by constantly asking questions that start with why, what, and how. "Why are you adding baking powder? Why are you not adding baking soda? How did you know how long to keep it in the oven? Why are there bubbles on my cookie?"

Although you are not very motivated to answer the questions of your sister, you still would like to earn some extra credit from your chemistry teacher and so decide to help your mom and practice your chemistry skills in the kitchen.

Good luck!

ENGAGEMENT

The teacher starts off the project by having an open discussion about the kinds of cookies students like, their role in family culture, and their benefits and possible harms to health. They also discuss the properties of a delicious cookie, how people in class like their favorite cookies baked differently (crispy, soft, etc.), and how companies producing different brands of cookies determine their recipes. The teacher encourages students from other cultures to share the role of cookies in their national cuisine and the different words they use for cookie: biscuits, *kels*, *galletas*, etc. After the discussion, students choose their favorite cookie from the list provided at http://www.popularcookierecipes.com/ and determine the most common ingredients used in their favorite cookie. They are then asked to write down the ingredients used and the instructions to bake it. This should take one class period.

EXPLORATION

During the next class meeting, the teacher presents the story in the Student Introduction above and asks students to find information about each of the ingredients used in their recipes and present their findings according to Rubric 1 (See Appendix 1). Information that the teacher may provide to enhance student interest and knowledge is given below (Czernohorsky, J. H. & Hooker, R., n.d.). This part of the exploration should take one class period.

Flour. Flour provides the bulk of the baked item. For bread baking, wheat flour, which is high in gluten (protein), should be used, as this is the substance that gives bread its fine texture and holds the other ingredients together during rising.

Yeast. Yeast is a fungus that feeds on starch and sugars, releasing CO_2, alcohol, and sugar. The CO_2 bubbles give the dough a light, airy texture and are responsible for the rising effect.

Fat. Fat gives a softer texture and helps prevent the CO_2 bubbles from escaping from the mixture too soon.

Sugar. Sugar provides a direct food source for the yeast, improving its action.

Baking Soda. Baking soda releases CO_2 upon encountering moisture and an acidic ingredient according to the equation:

$$NaHCO_3 + H^+ \rightarrow Na^+ + H_2O + CO_2 \qquad (1)$$

However, as this also produces a strong base, Na_2CO_3, which has to be neutralized, baking powder is usually used instead.

Baking Powder. Baking powder is baking soda with an acidifying agent added. This neutralizes the base and produces CO_2 the same as in the above equation but can be used in recipes without acidic ingredients.

Eggs. Beaten egg whites, like fat, help retain gas bubbles, whereas egg alone acts as a binder.

Salt. Salt adds flavor and strengthens soft fat and sugar mixtures.

The teacher opens another discussion about commercial cookie production and the necessary hygiene and health conditions that should be maintained at factories. Students are asked to investigate the role of the laboratory in commercial production and career opportunities in this field. They present their findings as an essay written according to Rubric 2. This discussion and investigation should take one class period.

EXPLANATION

The teacher takes students to the chemistry laboratory, and the students are assigned to a station with the following materials prepared beforehand: a pipette, Bunsen burner, triple beam balance, baking soda, vinegar, test tubes, and several rubber balloons. All chemistry lab safety procedures should be explained to the students before starting the experiment and observed throughout it. Students are expected to show and explain why baking powder is preferred over baking soda in cookies and what happens to the CO_2 byproduct when baking soda is used instead of baking powder. They should also be able to explain Avogadro's law as it is applied into Equation 1. The procedure will be straightforward to a chemistry teacher; students should heat the baking soda to activate $NaHCO_3$ (see Equation 1) and measure the amount of CO_2 released by measuring the weight of the balloons attached at the mouth of test tubes. In the second part of the experiment, students use baking soda with vinegar, which works as the acidic medium, and again measure the amount of CO_2 and compare. Students should be able to explain what happens to Na_2CO_3 in the acidic environment of a stomach, which results in the sour taste of cookies baked with baking soda instead of baking powder. Students go back to their classrooms and work on their lab reports, explaining their findings and their possible answers to the questions posed above. Rubric 3 is used to assess student effort and work. This lesson requires an approximately 90-minute period of class time for the experiment and another period for students to prepare their presentations.

EVALUATION

The rubrics are used to help students guide their learning. Rubrics are used as both summative and formative evaluations. Students have the experience of being rated both by someone using the rubrics as well as using the rubrics themselves. When they spend the time to rate each other, it improves their work. They should reflect on each rubric before it is used to provide formal feedback and grading both by someone using the rubric as well as by using the rubrics with each other (See Appendix A, B, C and D for sample rubrics).

EXTENSION

Start class by asking students if they have ever received a helium balloon for their birthday. After the students have answered, ask them whether their balloons suffered any changes. Did the level of inflation change? Can they recall anything that seemed to have an effect on the balloon, such as temperature or age? The extension activity is designed so that students can answer the following questions and show their work in a graphical format, possibly with the aid of a graphing calculator. Charles' Gas Law states that as temperature increases, volume increases (like the balloon). As temperature decreases, volume decreases, because volume and temperature are directly proportional. At a constant temperature, according to Boyle's Gas Law, as pressure increases, volume decreases; as pressure decreases, volume increases. Pressure and volume are also directly proportional. Students are given the following materials to show the relationship of the variables in Charles' Gas Law: two rubber balloons that are the same size but different colors (blue and red), a tape measure, two beakers, hot water, ice, and a thermometer. The teacher should not provide the formula for Charles' Gas Law or Boyle's Law but should rather expect students to come up with it according to the direct/indirect proportion relationships they have learned in algebra. After a revision of the formulas with the teacher, students can be asked to use the following terms to build a concept map that organizes the gas laws: Boyle's Law, Charles' Law, Avogadro's Law, pressure, volume, temperature, and number of moles.

APPENDIX A

Research about the chemistry of cooking

	RESEARCH	*PRESENTATION*	*GROUP WORK*
4–3	Practical benefits and the function of each ingredient are given using correct chemistry vocabulary. Students demonstrate a complete understanding of each chemistry term used, and students' notes in their explanations are their own words rather than a copy-paste from the resource. A variety of resources are used, and they are indicated separately.	Great visualization and a very clear presentation of each ingredient.	Each student worked diligently, and the group shared responsibilities according to each member's strength.
2–1	Some of the terms used are misunderstood or the explanations are not clear. Cited resources are not indicated clearly.	The visualization is poor. Some of the ingredients are not presented at the required level. The terms used are not explained clearly.	Some students did most of the work, whereas others made a minimal contribution.
0	No attempt or mostly copy-paste work.	No attempt or totally careless presentation.	All the work is done by certain students; others had no contribution.

APPENDIX B

Personal Essay Based on Research on the Commercial Cooking Industry

	RESEARCH & CONTENT	*WRITTEN WORK*
4–3	Thorough explanation of Chittick's Method and Atomic Absorption Spectroscopy in the student's own words and good examples of differences between home cooking and commercial cooking. Good explanation of the roles of different professionals working in various departments in a factory setting. Complete understanding of each chemistry term used. A variety of resources are used, and they are indicated separately.	The topic is clear, explanations are logical, there are no digressions, and transitions are smooth and logical. Clear sense of beginning and end. Great personal touch; the student's thoughts about working in the commercial cooking industry are included as well as the thoughts of others.
2–1	Some of the terms used are misunderstood or the explanations are not clear. There is no reference to Chittick's Method or Atomic Absorption Spectroscopy, and the explanation of the difference between home and commercial cooking is inadequate. Cited resources are not indicated clearly.	The topic is confusing, and there are some digressions or over-elaborations that interfere with reader understanding. No clear sense of beginning and end.
0	No attempt or mostly copy-paste work.	No attempt or much digression or over-elaborations that cause significant interference with reader understanding.
Comments		

APPENDIX C

Table 3. Rubric 3: Personal essay based on research on the commercial cooking industry

	Scientific Methodology	*Predictions*	*Facts*	*Applications*
5	All procedures and reporting show evidence that the scientific method was followed.	Predictions are relevant and all are accurate.	Facts are relevant and all are accurate.	Applications made are supported by the data and all are accurate.
4	Almost all procedures and reporting show evidence that the scientific method was followed.	Predictions are relevant and most are accurate.	Facts are relevant and most are accurate.	Applications made are supported by the data and most are accurate.
3	Most procedures and reporting show evidence that the scientific method was followed.	Predictions are relevant and some are accurate.	Facts are relevant and some are accurate.	Applications made are supported by the data and some are accurate.
2	Procedures and reporting show some evidence that the scientific method was followed.	Predictions are relevant but inaccurate.	Facts are relevant but inaccurate.	Applications made are supported by the data but are not accurate.
1	Procedures and reporting do not follow the scientific method.	Predictions are irrelevant and inaccurate.	Facts are irrelevant and inaccurate.	Applications made are not supported by the data and are not accurate.
0	No reporting or procedures are included.	No predictions are included.	No facts are reported.	No applications are made.
Comments				

Evaluation

	Scientific Methodology	*Predictions*	*Facts*	*Applications*
5	All procedures and reporting show evidence that the scientific method was followed.	Predictions are relevant and all are accurate.	Facts are relevant and all are accurate.	Applications made are supported by the data and all are accurate.
4	Almost all procedures and reporting show evidence that the scientific method was followed.	Predictions are relevant and most are accurate.	Facts are relevant and most are accurate.	Applications made are supported by the data and most are accurate.
3	Most procedures and reporting show evidence that the scientific method was followed.	Predictions are relevant and some are accurate.	Facts are relevant and some are accurate.	Applications made are supported by the data and some are accurate.
2	Procedures and reporting show some evidence that the scientific method was followed.	Predictions are relevant but inaccurate.	Facts are relevant but inaccurate.	Applications made are supported by the data but are not accurate.
1	Procedures and reporting do not follow the scientific method.	Predictions are irrelevant and inaccurate.	Facts are irrelevant and inaccurate.	Applications made are not supported by the data and are not accurate.
0	No reporting or procedures are included.	No predictions are included.	No facts are reported.	No applications are made.
Comments				

REFERENCES

Czernohorsky, J. H., & Hooker, R. (n.d.). Virginia Humphrey-Taylor. *The New Zealand Institute of Chemistry*. Retrieved June 11, 2009, from http://nzic.org.nz/ChemProcesses/food/6D.pdf

Verastequi, L. (n.d.). Gas Laws. *Texas Tides*. Retrieved June 5, 2009, from http://tides.sfasu.edu/teachers/ LessonPlans/LauraVerastegui/ VerastuguiGasLaws.pdf

M. Sencer Corlu
Aggie STEM Center,
Texas A&M University

Kathy Vondracek
Chemistry Teacher,
Emmett J. Conrad High School, Dallas ISD

CRYSTAL J. WOODS

23. CONRAD FOODS: CONSUMERS NEED A NEW FOOD PRODUCT AND AN ATTRACTIVE LABEL

WELL-DEFINED OUTCOME

Using Internet resources, students will design a label, describe the chemical composition, and calculate the nutritional value of a new food product.

INTRODUCTION

Students will design a food-product label that covers the lateral area of a cylindrically shaped object. Many food products are stored in cylindrical containers such as cans or bottles, and they have a variety of labels. Using labels available on the market as models, students will develop a label for a food product of their choice. In this project, students will experience the decisions needed to be made in the process of label development. As students work on this project, they will consider the attractiveness as well as the effectiveness of the labeling. Students will also chemically describe the composition of the food with an emphasis on the macromolecules involved and the nutritional value of the product. This project-based learning (PBL) activity follows an interdisciplinary approach and will enable students to integrate their knowledge from subjects such as language arts, mathematics, and science.

OBJECTIVES

Mathematics

- Students will estimate measurements and formulas to solve application problems involving lateral and total surface area and volume.
- Students will identify and apply mathematics to everyday experiences, to activities in and outside of school, with other disciplines, and with other mathematical topics.
- Students will use a problem-solving model that incorporates understanding the problem, making a plan, carrying out the plan, and evaluating the solution for reasonableness.

Language Arts

- Students will develop a clear thesis or position based on logical reasons supported by precise and relevant evidence.
- Students will develop counter-arguments based on evidence to anticipate and address objections.
- Students should provide an analysis of the relative value of specific data, facts, and ideas.
- Students should use graphics and illustrations to help explain concepts where appropriate.
- Students will use a variety of evaluative tools (e.g., self-made rubrics, peer reviews, teacher and expert evaluations) to examine the quality of the product.

Sciences

- Students will communicate and apply scientific information extracted from various sources such as current events, news reports, published journal articles, and marketing materials.
- Students will compare the structures and functions of different types of biomolecules, including carbohydrates, lipids, and proteins.

R. M. Capraro, M. M. Capraro, J. Morgan and J. Scheurich, (eds.), A Companion to Interdisciplinary STEM Project-Based Learning: For Teachers by Teachers, 207–212.

- Students will use calorimetry to calculate the heat of a chemical process.
- Students will illustrate the transformation of energy within an organism, such as the transfer of chemical energy to heat and thermal energy in digestion.

Graphic Communication Technologies

- Students will – Demonstrate knowledge of electronic and other image generation devices for reproduction photography.
- Students will – Demonstrate knowledge of new and emerging technologies that may affect the field of graphic communication technologies.
- Students will – Create communication materials utilizing color, text, and graphics.

MATERIALS NEEDED

- A cylindrically shaped object (e.g., soda can, canned foods)
- Paper nets patterned for the cylindrically shaped objects (for viewing purposes)
- A variety of material to design packaging (such as construction paper, wallpaper scraps, etc)
- Scissors
- Colored pencils or markers
- Glue
- Formula chart
- Centimeter grid paper
- Yarn
- Ruler
- Bunsen burners
- Food samples
- Beakers
- Thermometers
- Calorimeters
- Safety goggles
- Hot pads or oven mitts
- Optional PowerPoint presentation and/or visuals (extra credit should be awarded if a visual is provided)

STUDENT INTRODUCTION

Conrad Foods is a food company in town. To be able to survive in today's competitive marketplace, Conrad Foods is under pressure to develop not only innovative products but also attractive labels that catch the consumer eye. Thus, Conrad Foods is in search of a new label for a new innovative product that will attract consumers and get business booming again. Because teenagers constitute a large percentage of Conrad Foods' consumers, the CEO, Ms. Woods, has decided to get ideas from students at Conrad High School on the development of a new food product and its labeling in order to make it attractive to market. Thus, Ms. Woods wants you to create a new food product that can be packaged in a cylindrical shape with an attractive and effective label. You will also need to explain and prove mathematically the nutritional benefits of your product.

DAY 1

Engagement

The teacher engages students in a conversation about a company, Conrad Foods, whose sales are in a downward spiral. The CEO wants each team member to design a new product that will get business booming again. The teacher encourages students to discuss their ideas about what the new food product could be. Students can share their ideas for the product and their reasons. The teacher can, then, get the students to talk about what is on food labels and what makes the food attractive to the consumer. The teacher can write down students' ideas on a sticky tabs or a projector. This part of the project should take between 5–15 minutes.

Exploration

Students sketch and/or brainstorm ideas about what product they will produce using a cylindrically shaped object as well as about the things that must be included on the product label. Students can use magazines or the Internet to get ideas about attractive and effective food labels. The teacher draws a chart on the board as students create the same chart in each of their journals. The students will engage in a teacher-led discussion to fill in the chart. An example is provided below in Table 1.

Table 1. Sample journal entry

Characteristics of attractive labels	*Characteristics of effective labels*	*Characteristics of nutritious foods*
Colorful Neatly written Eye appealing	Easily visible Easily readable Interesting to the recipient Make people look	Low fat Vitamins and minerals Complex carbohydrates High protein content

The chart can help students compare the features of attractive and effective labels and the necessary nutritional requirements. Are all effective labels attractive or vice versa? Do all food labels have the same information on the label?

Brainstorming should result in a sketch that includes the product name, designs and/or logos, and the components that will be included on the label. During the exploration, the emphasis is on the label and the lateral area of the cylindrical shape. This part of the project should take 30 minutes, but some students may need more time for brainstorming. However, all of the students must have a rough draft ready to go by the next class period.

DAY 2

Explanation

Students bring cylindrical shapes of their choice to the class and are expected to start creating nets of flat patterns for their cylindrical shapes using centimeter paper, a ruler, yarn, scissors, and a formula chart. With given material, students must determine all of the dimensions of the shape such as the length, width, height, circumference, radius, and diameter so that they can compute the total and lateral surface area and volume of the cylinder. Students also need to create an illustration of the net flat pattern of the cylinder on a page of centimeter paper using the measurements collected. The illustration of the flat pattern as well as the flat pattern itself should be labeled with the respective measurements.

As students work on their measurements and flat patterns, they will be given the rubric that will be used to evaluate their products (see Appendix A). The teacher can use the rubric in the following ways: (a) The teacher and all the students will evaluate the presenter with the rubric. The teacher will assign a grade that is the average of the classmates' rubrics and the teacher's rubric, or (b) students will grade presenters with the rubric but only the teacher's rubric will count for a grade. Teachers are encouraged to discuss the content of the rubric and how the rubric should be used; they should discourage students from rating their peers based on personal relationships. See the Appendix B for a sample rubric. This explanation should take between 45 minutes to an hour.

DAY 3

Using sketches of their labels, students will create actual labels for their food products with any material (mostly with construction paper) that will cover the lateral area or the total area of their cylindrical shapes (see pictures in the Student Example section). Thus, students bring their flat patterns of the cylindrical shapes to life. Students attach the new labels to the lateral area (or the total area) of the cylindrical shapes using any material.

Presentation preparations may begin at this time. The teacher advises students that each student will present his or her product to the class. Presentations have to be persuasive, and students must convince the other students that their product should be marketed. All dimensional information (dimensions, volume, and lateral and total surface area) must be included in the presentation as well. Students should be given 45 to 60 minutes for these activities; if necessart, the rest can be class work or homework, depending on time constraints.

EVALUATION

Presentations

Students will present their newly designed products evaluated based on the presentation rubric in Appendix B. Teachers should make sure students are questioning presenters with appropriate questions. Presentations must have some type of visual aid (chart, graph, powerpoint are some examples). Presenters should justify all calculations (surface areas and volumes) by connecting the shapes on the flat patterns to the actual 3D calculations. Students should see that the flat pattern calculations should be equivalent to the 3D calculations. Students should understand that the areas of the separate shapes are congruent to the surface areas. Students who are evaluating should probe presenters about why they should choose their product. The size of the class will determine the presentation time allotted to each student, which will probably be between 3–5 minutes per student.

DAY 4 TO DAY 5

Extension

In their journals, students will take their created label and analyze the food for the macromolecules included, the caloric content, and the overall nutritional value of their food. Students will generate a graph of their choice that shows this information. Then students will calculate the total calories using dimensional analysis and energy content of their food using the formula $Q = (m)(\Delta T)(C_p)$. Student groups will then compare the energy capacity of their foods using calorimetry to demonstrate the amount of heat eliminated from their food product. If the food products chosen are not actual products, students may choose a food that has a similar macromolecule and caloric content. Students will record all data and observations in their journals and will write at least two paragraphs explaining and relating their findings to real life challenges such as diet and weight loss. Assessment will follow the checklist found in the Appendix C.

APPENDIX A

Project Rubric

Name:______________________________ Teacher:______________________________

Date Submitted:________________________ Title of Work:_________________________

Criteria	*4*	*3*	*2*	*1*	*Pts.*
Explanation	A complete response with a detailed explanation.	Good, solid response with clear explanation.	Explanation is unclear.	Misses key points.	
Use of Visuals	Clear diagram or sketch with substantial detail.	Clear diagram or sketch.	Inappropriate or unclear diagram or sketch.	No diagram or sketch.	
Mechanics	No mathematical errors and clear and logical reasoning.	Some minor mathematical errors or flaws in reasoning.	Major mathematical errors or flaws in reasoning.	Does not make sense mathematically.	
Demonstrated Knowledge	Complete understanding of the questions, mathematical ideas, and processes.	Substantial understanding of the questions, mathematical ideas, and processes.	Some understanding of the problem.	Lack of understanding of the problem.	
Requirements	Goes beyond the requirements of the problem.	Meets the requirements of the problem.	Hardly meets the requirements of the problem.	Does not meet the requirements of the problem.	
				Total	
Teacher's Comments					

APPENDIX B

Presentation Rubric

Name:______________________________ Teacher:______________________________

Date Submitted:________________________ Title of Work:_________________________

Points	*RESEARCH*	*PRESENTATION*	*Total*
4–3	Thorough investigation of labels on the market, which includes the reasons behind the companies' choices of particular labels. A variety of resources are used, and they are indicated separately.	Great visualization supported by maps and figures. Knows the content of the presentation, and it flows logically.	
2–1	Mediocre research of the labels available on the market. Lack of substantiation for selection of labels. Resources are not indicated clearly.	Some visualizations; unclear presentation of the topics discussed in the research. The student is somewhat familiar with the content of the presentation.	
0	No attempt or mostly copy and paste work. Unable to discuss the topic and lack of rationale for choices and decisions.	No attempt or poorly conceptualized presentation; visuals may be lacking or inappropriate for the content.	

APPENDIX C

Extension Checklist

CATEGORY	*Evident (1–4/5)*	*Not evident (0/5)*	*Points*
Macromolecule content is described.			
Macromolecule content is demonstrated in graph format.			
Caloric content and heat capacity are calculated.			
Calorimetry experiment data is collected.			
Conclusion relates food content to diet and food choices.			
		TOTAL POINTS	/ 25

STUDENT EXAMPLES

Crystal J. Woods
Mathematics Teacher,
Emmett J. Conrad High School, Dallas ISD

NEVRA CETINER AND MARY MARGARET CAPRARO

24. SELLING HAND-MADE ITEMS IN A TURKISH SCHOOL KERMES*

WELL-DEFINED OUTCOME

Students will organize a Kermes* (a market composed of stands and set up in Turkish schools for a few days for the purpose of raising money for charity) in order to benefit their school library and apply their mathematical knowledge in a real-life activity immediately after learning it. The mathematical content will make sense to them.

INTRODUCTION/BACKGROUND

The teacher will present different types of graphical representations to the class. Students are encouraged to discuss and brainstorm about the graphs. The following questions and possible answers may be useful for this process:

What are the components of a graph?. Students need to investigate examples of graphs in order to answer this question. "Variables" and "a system that shows values" are examples of satisfactory answers.

When should graphs be used?. Students need to think about the benefits of graphs in order to be motivated to learn the subject and content.

What must be known in order to create or interpret a graph?. After investigating graphs, students should determine the prerequisites of the content themselves. In this way, they will figure out how percentage calculations are connected to interpretations of graphs.

What is a graph?. Students need to organize and construct the meaning of different types of graphs in their mind and be able to explain them clearly to others.

OBJECTIVES

Social Science

- The student creates thematic graphs, charts, models, and databases representing various aspects of the world.
- The student poses and answers questions about patterns shown on graphs, charts, models, and databases.
- The student understands how physical and human characteristics of place can change.

Mathematics

- The student will be able to interpret graphs based on variables.
- The student will be able to create graphs according to data they collect themselves.
- The student will be able to make decisions about real-life situations according to statistical representations.
- The student will be able to represent relationships among quantities using concrete models, graphs, and diagrams.

Language Arts

- The student uses prewriting strategies to generate ideas.
- The student compiles written ideas and representations into reports, summaries, or other formats and draws conclusions.

R. M. Capraro, M. M. Capraro, J. Morgan and J. Scheurich, (eds.), A Companion to Interdisciplinary STEM Project-Based Learning: For Teachers by Teachers, 213–219.

Possible Social Outcomes for Students

- Students working in groups will develop communication skills and experience the fun of cooperation and collaboration.
- Students will have the chance to meet their teachers in an atmosphere other than the classroom.

Possible Social Outcomes for Teachers and Parents

- Teachers and parents will have a chance to meet in a positive atmosphere to see student work.
- Teachers will have a chance to learn more about the family background of students.
- Parents will learn about their children's successes in school.

MATERIALS NEEDED

During the Lesson

- Examples of graphs to study in the classroom

On Kermes Day

- PowerPoint slides to present the preparation process
- Specific items determined by students in order to set up a stand (table, table cover, etc.)
- Posters and slogans prepared by students that will increase sales

Time frame for the project. This project will take place over five days, including the extension activities.

DAY 1

Engagement

Students should be arranged into groups for each stand; throughout the project, they work in these same groups. Each group will decide what items (food, handmade products, toys, etc.) they will sell at the Kermes. They may, if necessary, ask for help from their parents for this step. Students decide the prices of their products. To do this, they should calculate the cost of products. For example, if they want to sell apple pie in the Kermes, they will check the prices of flour, eggs, apples, oil, etc. at markets. Next, they establish a graph of different interest rates for the cost of products. Then they will determine the most appropriate price for their products. Each group is encouraged to design an advertisement strategy with tools such as posters, slogans, etc. On Kermes Day, each group will set up a stand and sell items to parents, friends, and teachers.

DAY 2

Exploration

Interpreting various graphs. In this step, various graphs are presented to the groups, and each group must interpret the graphs and present an interpretation to the class. The following examples of graphs (see Figures 1, 2 and 3) may be useful:

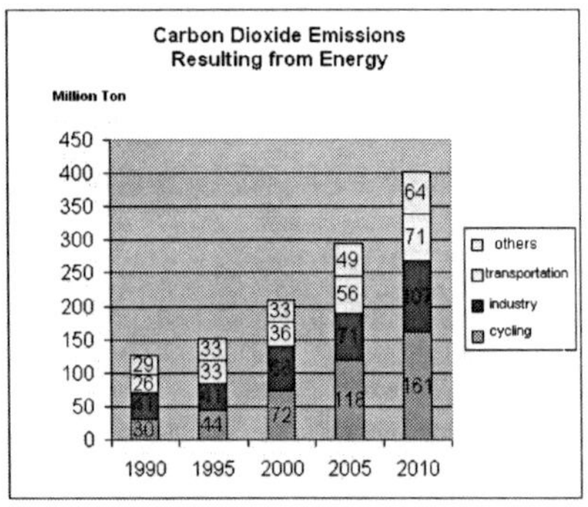

Figure 1. Carbon dioxide emissions resulting from energy.

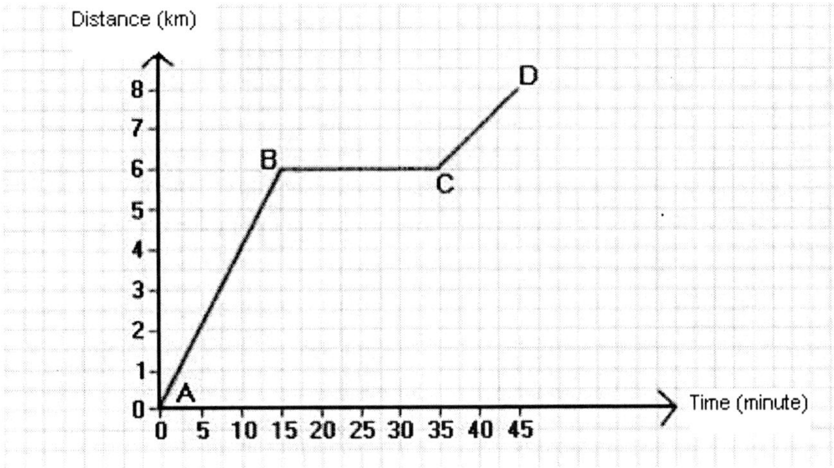

Figure 2. The distance a bike rider covers vs. time.

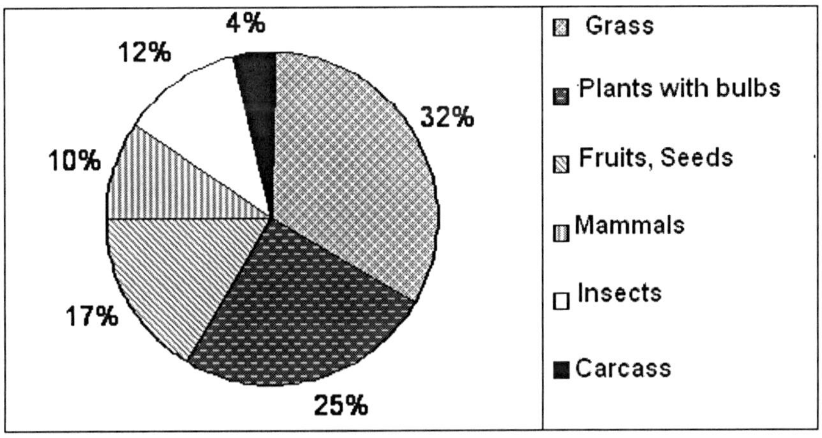

Figure 3. Foods grizzly bears eat in national parks.

DAY 3

Explanation

Before working in groups, students need to know what questions they are supposed to answer during their presentations. The following questions may be used:

- What does your graph show?
- How did you figure out that you needed to use that type of graph?
- What are the variables used in your graph?
- How do they affect each other?

Teachers can ask groups more "how" and "why" questions while walking around. Students should try to prove their answers and look for reasons and justifications.

DAY 4

Students should use the data they collected during Day 1 to calculate the cost of products and appropriate sale prices. Finally, they should create representations (graphs or tables) to organize, record, and communicate their calculations. Students should understand the goal, which is to achieve a reasonable profit. It may be

Table 1. Apple pie ingredient prices

Ingredients	***Prices (TL*)***
1 kg flour	1.10 TL
250 gr sugar	0.7 TL
2 eggs	0.4 TL
2 apples	0.6 TL
250 ml fat	1.20 TL
Total cost:	**4 TL**

Table 2. Possible profits based on apple pie prices

Rate of Profit (%)	*Profit (TL)*	*Price of the apple pie (TL)*	*Price for 1 slice (TL)*
25	25% * 4 = 1	4 + 1 = 5	0.625
50	50% * 4 = 2	4 + 2 = 7	0.875
100	100% * 4 = 4	4 + 4 = 8	1
200	200% * 4 = 2	4 + 8 = 12	1.5

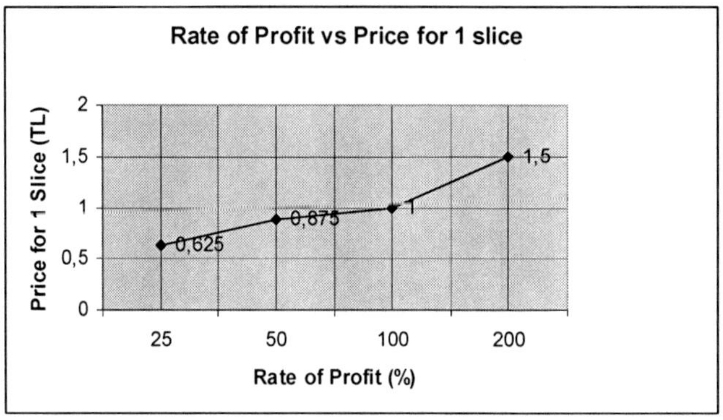

Figure 4. Various rates of profit for apple pie.

better if students state the variables of the graphs or at least suggest some options. The teacher can write some options on the board after a brief discussion and brainstorming session such as the ones listed here:

- Graphs of prices based on different profit rates.
- Graphs of the prices of products' ingredients (circle graphs are recommended).
- Graphs of price-increase based on profit rates.

An example is provided:

Group A decided to sell apple pie at its stand at the market on Kermes Day. They calculated the cost of an apple pie divided into eight pieces. The prices for apple pie ingredients given in Table 1 are in Turkish Liras (TL). Table 2 provides possible prices for pieces of apple pie depending upon how much profit students feel is reasonable. Figure 4 provides the various rates of profit depending upon prices of apple pie slices.

DAY 5

As a culmination to the project, all groups will present the strategies they used for determining the sale prices of their products. In their presentations, the teacher should require that students use tables and graphs in order to encourage mathematical thinking, In the discussion part, all groups will analyze and evaluate the mathematical thinking and strategies of each of the other groups using the rubric located in Appendix A. Students may evaluate the strategies used for determining the sale prices. They may find prices highly expensive or cheap. Students need to provide mathematical reasons to support their critiques.

They may claim that profit rates are too high or too low and suggest different rates with an appropriate rationale. During this part of the project, students should communicate their mathematical thinking clearly with others and use mathematical representations for communication.

EXAMPLES OF STUDENT WORK

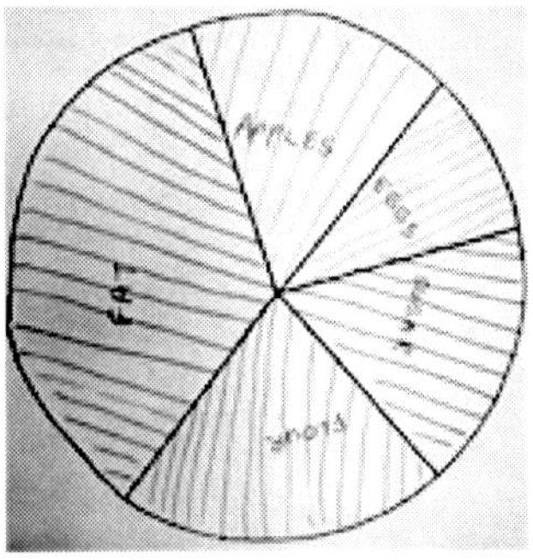

Figure 5. Pie chart of products for apple pie ingredients.

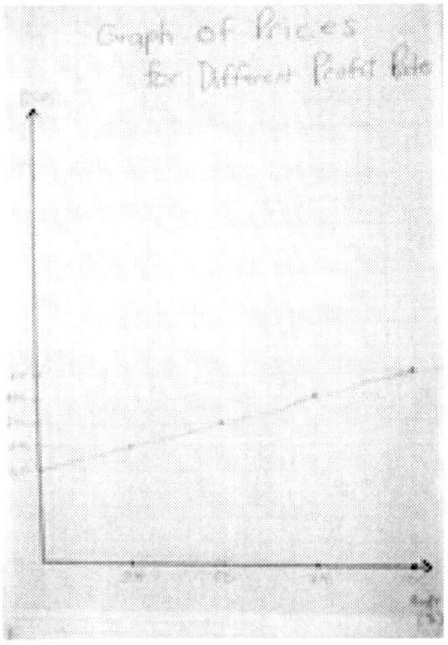

Figure 6. Graph of pieces for different profit rates.

Reflection. During the project, students should apply their mathematical knowledge to real-life situations. With this project, teachers can lead students to build mathematical knowledge through problem-solving strategies. When they are faced with real-life problems, they need to apply and adapt a variety of appropriate strategies to solve them. Because they are building their own mathematical knowledge, this knowledge will be meaningful and permanently learned. Students will retain this knowledge and be able to use it when selling their products on Kermes Day.

Extensions

A field trip to a market or Grand Bazaar may be organized to help students make decisions about what they will sell on their stand. They can collect data about products and then determine reasonable prices for selling their items at the school Kermes Day.

Kermes Day. Students make posters and determine slogans that show marketing strategies in order to contribute to their sales. After the preparation process, parents and other teachers will be invited to Kermes Day. Students will present their sale items, slogans, and mathematical reasoning processes including their graphs to all guests by charts or PowerPoint slides. Then, they sell their products at the end of the Kermes school day. The rubric in Appendix B is a tool to evaluate their sales strategies.

APPENDIX A

Rubric for Using Graphs for Determining Sale Prices

Student Name: ______________________________

	4	*3*	*2*	*1*
Mathema-tical Concepts	Explanation showed a correct understanding of graphs based on the variables used to solve the problem(s).	Explanation showed some understanding of graphs based on the variables used to solve the problem(s).	Explanation showed a limited understanding of graphs based on the variables used to solve the problem(s).	Explanation showed an incorrect understanding of graphs based on the variables used to solve the problem(s) OR is not written.
Strategy Procedures	Used an efficient and effective strategy to determine the prices for the market.	Mostly used an effective strategy to determine the prices for the market.	Sometimes used an effective strategy to determine the prices for the market but did not do so consistently.	Rarely used an effective strategy to determine the prices for the market.
Diagrams and Sketches	Diagrams and/or sketches were clear and greatly added to the reader's understanding of the price determination procedure(s).	Diagrams and/or sketches were mostly clear and easy to understand.	Diagrams and/or sketches were somewhat difficult to understand.	Diagrams and/or sketches were difficult to understand or were not used.
TOTALS				

APPENDIX B

Rubric for Selling an Item at the Market on Kermes Day

Student Name: ______________________________

	4 points	*3 points*	*2 points*	*1 point*
Neatness and Organizationof Items for Sale	Items were presented in an attractive manner; they looked pleasing and were organized for selling.	Items were presented in a mostly attractive manner; they mostly looked pleasing and were mostly organized for selling.	Items were presented in a fairly attractive manner; they looked somewhat pleasing and were somewhat organized for selling.	Items were sloppily presented and disorganized for selling.
Cooperation When Working with Others to Prepare the Items for Sale	Student was an engaged partner who listened to the suggestions of others and worked cooperatively throughout project.	Student was an engaged partner but had trouble listening to others and/or working cooperatively.	Student cooperated with others, but needed prompting to stay on-task.	Student did not engage effectively and cooperatively with others.
TOTALS				

REFERENCES

Carbon dioxide emissions resulting from energy. (2009). Retrieved on July 1, 2009, from http://www.iklim.cevreorman.gov.tr/Gazi/grafikler2.htm.

Distance a bike rider covers vs. time. (2009). Retrieved on July 2, 2009, from www.matematikcifatih. tr.gg/grafikler.htm.

Foods that grizzly bears eat in a national park. (2009). 7th grade mathematics curriculum. Ankara, TK: Ministry of National Education.

Van de Walle, J. A. (2007). *Elementary and middle school mathematics*. Boston: Pearson.

Nevra Cetiner
Preservice Teacher Candidate,
Bogazici University, Istanbul, Turkey

Mary Margaret Capraro
Teaching, Learning, and Culture,
Texas A&M University

HIGH STAKES QUESTIONS

LISA R.W. COBB

25. PROJECT BASED LEARNING ASSESSMENTS

Project Based Learning (PBL) experiences often require a heterogeneous conglomeration of assessments. The nature of the PBL lends itself uniquely to differentiated instruction; because of this, the teacher has the opportunity to use multiple platforms to assess the student's learning. The varied assessments that accompany PBLs are especially suited for the diverse and multi-level population of students we have today.

There are two major types of assessments that are utilized in PBLs. During the course of the project, formative assessments evaluate student learning and progress as they move through the project. Journaling is an excellent tool for keeping track of learning during the phases of the PBL. Journal entries can be graded following a rubric or grading checklist. A product or model may be produced during the exploration, explanation, or extension phases. These may also be evaluated with a rubric. These are just a few examples of formative assessments. All assessment pieces should be evaluated with a rubric that is provided to the students ahead of time to communicate clearly what is expected.

The second major type of assessment is the summative assessment. The summative assessment is meant to evaluate if the student has achieved the well-defined outcome and the given objectives that accompany the PBL. Often multimedia products are used in a summative fashion. Examples include PowerPoint, videos, commercials, or brochures. These multimedia products may need to be communicated through a physical presentation component. These summative assessments should also be evaluated through the use of rubrics.

The more traditional summative assessment is the paper and pencil test, often composed of multiple-choice questions, such as our state assessments. Included in this chapter are examples of multiple-choice questions that could be used as summative assessment items for the respective chapters. There are questions that assess content knowledge, comprehension, application, and evaluation. All levels of learning must be assessed to ensure that students have achieved mastery of the desired objectives. This requires that the questions be carefully developed to ensure the variability of the thinking level as well as alignment to the objectives of the PBL. When developing questions, a team approach is best to make sure that questions are appropriate and clear in language and format. Lastly, Question writers should always have a second person review the content to ensure viability.

NOTE: The numbers of the questions correspond to chapters in the book. Thus Questions 1a,1b, etc all are assessment items that can be used to assess student's knowledge of the PBL in chapter one on earthquakes. High-stakes questions for ten chapters are included here.

1a. The theory that Earth's crust is divided into slow-moving plates that are responsible for forming various landforms is

A abyssal plains

B rift valleys

C* plate tectonics

D tectonic forces

1b. __________________ activity describes how plates slowly move in the upper mantle, which is called continental drift. At the boundaries of these plates the crust may melt, break, or bend causing volcanoes, __________________, and ____________________.

A core, magma, mantle

B tectonic, weathering, erosion

R. M. Capraro, M. M. Capraro, J. Morgan and J. Scheurich, (eds.), A Companion to Interdisciplinary STEM Project-Based Learning: For Teachers by Teachers, 223–236.

C earthquake, trenches, valleys

D* tectonic, folds, faults

1c. Some places are more prone to earthquakes and volcanoes. What factors enhance the susceptibility of a location to these forces?

A* the location lies along a huge fault line

B it has mountains made by tectonic plate movement

C it is located far out in the ocean

D it has excessive precipitation, causing weathering and erosion

1d. Islands in the Pacific Ocean are particularly susceptible to volcanoes and earthquakes. Which of the following explains why?

A the temperate climate

B their ancient culture

C* the location in the Ring of Fire

D the isolation from the mainland

1e. A student decides to study earthquake activity. What would be the best source for valid information about this topic?

A National Audubon Society

B Natural Resources Conservation Service

C* U.S. Geological Survey

D U.S. Naval Observatory

1f. Where and when did the largest earthquake occur in the 20th century?

A Northridge, California earthquake, 1994

B San Francisco, California earthquake, 1906

C* Chilean earthquake, South America, 1960

D Kobe, Japan, 1995

1g. A student is making a concept map about the different spheres that make up the earth and the phenomena that take place in the various spheres. Within what sphere do earthquakes occur?

A Atmosphere

B Biosphere

C Hydrosphere

D* Lithosphere

1h. Scientists hope to reduce damage by predicting where and when earthquakes will occur. They attempt to predict earthquakes by

A studying lava levels in the ground

B evaluating an area's weather patterns

C* studying an area's tremor history

D measuring fault depth in the ground

1i. Landforms such as plateaus are sometimes the result of

A climatic changes

B gravitational pull

C* tectonic movement

D atmospheric pressure

1j. Within what depth range do the majority of earthquakes occur?

A deeper than 700 km

B 700 km – 300 km

C 300 km – 70 km

D* 70 km – 0 km

3a. People affect their environment in various ways. Which of the following would have the LEAST affect on how people interact with the environment?

A cultural values

B political freedom

C* resources available

D technological advances

3b. Human actions have led to many effects on the environment, both positive and negative. Which of the following contributed to the problem of acid rain?

A Nuclear testing in desert areas

B Use of fertilizers on farmland

C* Factories burning fuels

D Release of natural gas from mining

3c. One negative impact that the use of modern technology is that

A contacts with other cultures decrease

B economic opportunities are limited

C* levels of air pollution often increase

D industrialization is adversely affected

3d.

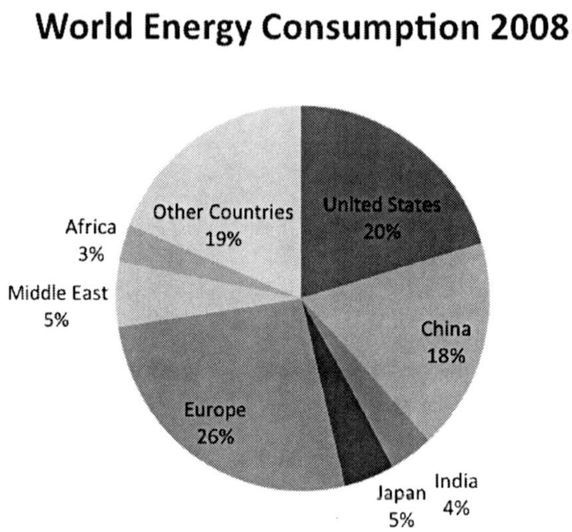

A Europe uses the most energy in the world

B Japan and the Middle East use the least of the world's energy

C* United States and China use more than 1/3 of the world's energy

D India and Africa use ¼ of the world's energy

3e. What is the one way the diesel motor has affected the physical environment?

A Its engines burn coal, which is a limited resource

B Its engine uses ethanol, which is a limited resource

C Its fuel contains more pollutants than any type of fuel

D* Its strength makes it important in earth-moving projects

3f. All of the following are reasons to protect watersheds except

A They provide drinking water.

B* They promote pollution.

C They sustain wild life.

D They provide recreation.

3g.

Global Warming	
Perspective 1	Contrasting Perspective
Human activity, including burning fossil fuels, increases the amount of carbon dioxide in the atmosphere. The more carbon dioxide in the air, the more heat is absorbed by the atmosphere which results in a rise in global temperatures.	?

Which of the following demonstrates a contrasting perspective?

A Scientists believe that carbon dioxide levels will double in the next 100 years. These levels will cause Earth's temperature to rise about three degrees. A warmer climate could cause a variety of problems, such as a rise in sea level and severe droughts.

B Carbon dioxide is one of the gases that allow Earth to absorb and retain heat in the atmosphere. Otherwise the heat would be released into space. This warmth allows plants and animals to survive on Earth.

C Industrial societies are dependent on coal, oil, and natural gas. Burning these resources causes an increase in the amount of carbon dioxide that is released into the atmosphere. Consequently, people are completely to blame for global warming.

D* Earth experiences natural cycles of warming and cooling. Climate changes globally over thousands of years. Because climate records have been kept for less than 200 years, it is impossible to determine the effect of human activity.

3h. Which of the following is a result of steps taken by humans to adapt to their environment?

A Antipollution devices

B Command economies

C* Desalinization plants

D Political boundaries

3i. In the past, wetlands were drained and filled to make farmland. What do most environmentalists suggest we should do with wetlands today?

A Maintain some and turn others into farmland

B* Preserve the wetlands and their vital ecosystems

C Continue to turn wetlands into farmland

D Convert some farmland into new wetlands

3j. All of the following are methods for reducing the human effect on the environment except

A Utilize composting, better irrigation systems and plants native to the specific climate

B Collect and recycle objects that can be recycled and reused

C Use water conservation devices and practices

D* Watching the level of energy usage to make sure it does not increase

10a. To determine the voltage available in a circuit containing batteries in series

A add the reciprocal of the voltage of each battery

B subtract the reciprocal of the voltage of each battery

C* add the voltage together for each battery

D multiply the voltage of each battery

10b. One of the biggest differences between a series and parallel circuit occurs in the

A voltage of the circuit

B current in the circuit

C the voltage and the current

D* the resistance of the circuit

10c. Algebraically rearrange the equation for the total resistance of three resistors connected in series, $R = R_1 + R_2 + R_3$ to solve for R_1 in terms of the other variables.

A $R_1 = R + R_2 + R_3$

B $R_2 = R + R_1 + R_3$

C* $R_1 = R - R_2 - R_3$

D $R = R_1 - R_2 - R_3$

10d. In a series circuit, how does the current compare in the different components of the circuit

A* the current is the same in all components

B the current is half as much in the first component

C the current adds up across all the components

D the sum of the reciprocal current in each component

10e. Ohm's Law mathematically relates what quantities?

A voltage, current, resonance

B voltage and current

C current and resistance

D* voltage, current, resistance

10f. Which of the following is NOT true about the rate of using energy, or the power available in a given circuit

A $P = IV$

B* $P = VR$

C $P = I^2R$

D $P = V^2/ R$

10g. John is making an electric car. He is using a nine volt battery and a motor that has a resistance of 4 Ohms. How much power can he generate with his car?

A 324 Watts

B 36 Watts

C* 20.25 Watts

D 144 Watts

10h. What data must a student collect in order to calculate the speed of a car?

A acceleration and time

B distance only

C* distance and time

D velocity and time

10i. When did exploration of Antarctica begin to be possible due to advancing technology?

A 1850s

B* 1920s

C 1950s

D 1970s

10j. Technology that made underwater exploration possible include all of the following except

A deep ocean submersibles

B untethered diving advances

C* microscope technology

D remote observation instruments

11a. Which of the following migrations was not a result of push factors?

A Sudanese Christians moving into Darfur

B European scientists coming to America during the 1940s–1950s

C Irish peoples moving to America in the 1800s

D* African slaves brought to America from the Caribbean

11b. Human migration is often a result of several motivating factors. Which of the following is a reason for people to choose a particular destination?

A civil war

B* low crime rate

C political strife

D religious oppression

11c. Settlers migrating to new lands often need to modify the land to suit their needs. Which of the following would help to modify the geography of the land to make a better lifestyle for the settlers?

A medical advances

B drought

C erosion

D* terrace farming

11d. All of the following are positive consequences of modern technology except

A contacts with other cultures increase

B economic opportunities increase

C industrialization increases

D* levels of pollution increase

11e. In the late 1800s, new farming technologies led people to view the Great Plains as a

A prime area for mining

B* vast area to be settled

C worthless dust bowl

D important industrial center

11f. Over time, perceptions of the American southwest have changed. Of the following descriptions, which best describes the past and present southwest?

A Past: farming and ranching
Present: dry, barren, and worthless

B* Past: wasteland with little life
Present: diverse landscapes, plant and animal life

C Past: location of large cities
Present: small towns due to lack of resources

D Past: tourism and recreation
Present: place to avoid due to extreme temperatures

11g. How did inventions such as the cotton gin, steel plow, steamboat, and telegraph affect the economy of the United States?

A* Increased profits and productivity

B Increased dependence on labor unions

C Increased numbers of small farms

D Increased westward travel by farmers

11h. Which of the following reflects the concepts of Manifest Destiny?

A Monroe Doctrine

B Expansion of slavery

C* Annexation of Texas

D Missouri Compromise

11i. The settlement of the American West was made easier by inventions and innovations developed before and after settling the "wild" west. All of the following factors had to be considered in order to improve old items or invent new objects or processes except

A* the physical attributes of the object

B availability of materials for construction

C the cost to manufacture the item

D durability of the final product

11j. What effect did advances such as the textile mill and the sewing machine have on women of the 1800s?

A improved pay scales

B expanded consumer credit

C* increased leisure time

D transformed agricultural labor

14a. All the continents on Earth have different landforms that characterize their physical surface. Which of the following continents does not have deserts?

A Africa

B Australia

C* Europe

D South America

14b. What is one benefit of using a political map instead of an aerial photograph?

A It shows distinct local vegetation

B It shows the layout of the land

C* It includes distance between locations

D It includes a three dimensional representation

14c. Which scale on map would provide you the most detail?

A 1: 30,000,000

B 1: 3,000,000

C 1: 300, 000

D* 1: 30,000

14d. All of the following affect the location of cities except

A latitude

B trade routes

C immigration patterns

D* traditions

14e. The landforms, soils, climates, and water of an area are known as its

A flora and fauna

B biodiversity

C* physical features

D context

14f. Which continent has the most regular coastline?

A Europe

B* Africa

C Australia

D Asia

14g. Which continent is the most isolated from the rest of the world?

A Europe

B Africa

C* Australia

D Asia

14h. Which two continents, due to their regular coastline and difficult navigation of rivers, have achieved less progress and discouraged European settlement of their land?

A Asia and Australia

B North America and South America

C Antarctica and Australia

D* Africa and South America

14i. Of the following factors, all have encouraged the development of human settlements except

A* elevation

B access to water

C natural resources

D climate

14j. Which continent is the best suited to contain the most advanced nations?

A Asia

B* North America

C Australia

D South America

20a. All of the following components make a gossip column different from a news article except

A short, snappy phrases

B puns, plays on words

C humor and intrigue

D* factual accounts

20b. Which of the following is a thesis sentence from a persuasive essay?

A We wept for wave orphans

B Although both wrote about their traumatic experience, the way each depicted the incident was distinct and seemed to be told for very different purposes.

C William Butler Yeats wrote two poems which are together known as the Byzantium series.

D* Antipersonnel landmines kill thousands of people every year.

20c. Quotations in a tabloid article are usually used to do all of the following except

A* communicate factual statements

B make an opening impact on the reader

C add humor

D closing leaving a point to think about

20d. When interviewing a person for an article, it is important to

A Talk quickly; they have a lot to do.

B* Prepare your questions ahead of time

C Try to remember everything without writing it down.

D Make a lot of jokes so the person thinks you are funny.

20e. Which of the following show the correct use of quotations?

A "My favorite part of the movie was the car chase." Andy said.

B Angel said. "that she wanted to go home."

C "Good luck," mom said!

D* "The person I most admire is my mom because she has always taken care of me."

20f. Which of the following topics is appropriate for a tabloid article about school?

A Mrs. Johnson meets Puff the Magic Dragon

B Coach Love Loves the Students

C Ms. Remington teaches good shooting technique

D* Ms. Travolta's Dancing Fever

20g. Which sentence provides a clue about the character's internal conflict?

A "You don't like people, do you?" he said sharply.

B "You're a very stupid person," she said.

C He had never talked to her so brokenly, and his sincerity, the depth of his feeling, began to stir her.

D* He longed to tell her that he was sure she would have been charming to his father, but he had never felt so wretched in his life.

20h. What change should be made in the following sentence: "At the age of 13, Spielberg made a 40-minute war movie that featured a battle between Nazi and British soldiers."

A Change ***made*** to **makes**

B Change ***British*** to **british**

C Change ***soldiers*** to **soldier's**

D* Make no change

20i. What change needs to be made in the following sentence: "The familar city smells were missing, too, such as the delicious aromas from the Portuguese, Mexican, and Thai restaurants on Front Street."

A* Change ***familar*** to **familiar**

B Change ***were missing*** to **are missing**

C Change ***too*** to **to**

D Change ***Street*** to **street**

20j. Which of the following is not appropriate for an informative essay?

A Think about a problem that you solved, and then write about how you solved the problem.

B Write about things that motivate people to do their best and provide specific examples.

C Write about these different roles that teenagers play and provide specific examples.

D* Write about who you think should be president and give several reasons to support your point of view.

22a. All of the following are examples of chemical changes except

A rising of bread

B* combining ingredients

C cooking a casserole

D sautéing vegetables and rice

22b. Which of the following is not responsible for the leavening, or rising, of baked goods?

A yeast

B* brown sugar

C baking soda

D baking powder

22c. How are baking cookies and the Gas Laws related?

A The cookies must be baked in air, which is a gas.

B The Gas Laws tell how much gas is needed to bake cookies.

C The Gas Laws are not related to baking cookies, only Charles' Law.

D* The ingredients and baking temperature relate to the Gas Laws.

22d. In order for a substance to be classified as an acid, it must

A* release hydrogen ions in solutions

B release hydroxide ions in solutions

C combine hydrogen and hydroxide ions

D use water to release any ions into solution

22e. Dimensional analysis relates to cooking for the following reasons except

A ingredients must be measured in a given proportion

B often conversions must be made from metric to conventional units

C substitutions or batch size changes can be made

D* making the cookies different sizes

22f. Which Gas Law states that in the presence of a constant temperature, the pressure and volume are directly proportional?

A* Boyle's Law

B Charles' Law

C Gay-Lussac's Law

D Avogadro's Law

22g. Acids react with bases to form

A hydrogen and oxygen

B nonmetal oxides and water

C* salts and water

D nonmetal hydroxides

22h. In order to properly balance a chemical equation, what are the only numbers that should be added or changed?

A* coefficients

B subscripts

C molar masses

D superscripts

22i. Which of the following is not evidence of a chemical reaction?

A* change in color

B heat and light

C formation of a precipitate

D formation of a gas

22j. Which of the following represents the reaction between hydrochloric acid and sodium hydroxide?

A $2HCl + Na(OH)_2 \rightarrow 2H_2 + Na^+ + Cl^- + O_2$

B* $HCl + NaOH \rightarrow H_2O + NaCl$

C $HCl_2 + 2Na(OH)_2 \rightarrow 2H_2O + 2NaCl + OH^-$

D $2HCl + 2NaOH \rightarrow Na(OH)_2 + H_2Cl_2$

JENNIFER STIT

26. CLOSING COMMENTS: REFLECTIONS OF STEM PBL TEACHER

MAKING EDUCATION MEANINGFUL

As educators, how do we make education meaningful and keep our students engaged? Society has evolved dramatically over the last 50 years due to technology. With time comes change and with change comes opportunity. Not only do we see changes in people and our economy, we see dramatic changes in our education system thus an opportunity to enhance our educational system. Educators have a wonderful array of backgrounds, ages and styles. New and seasoned educators use lecture and notes as the primary tool for presenting material. This style for presenting material is an efficient way to communicate an enormous amount of information in a short period of time, but does it make education meaningful? When you reflect on your education do you find meaning in the worksheets you filled out, or do you find meaning in the thought provoking projects you were a part of ?

Modern education competes with the internet, social networking, cell phones, and personal digital assistants, so teachers need to embrace and use these technologies to our advantage. How do we make education relevant, quick, up-to-date, in high definition, and interesting? How do we show mathematics and science through technology? These questions need to define education today not postulate about it. We need a future of young people who can go beyond wrote summary knowledge and more able to apply what they learned. Multiple choice tests and worksheets are not the most efficient way to retain information for application purposes. Project based, hands-on-learning is!

MAXIMIZING STEM EDUCATION?

The Science, Technology, Engineering, and Mathematics (STEM) education is a direct conduit into project-based learning (PBL). STEM education has many tools in its belt, and Geographic Information Systems (GIS) is one of the tools that mathematicians, scientists, and engineers alike use to help solve problems. Prior to entering the education profession three years ago, I was in industry using GIS software. GIS is the technology associated with maps and databases. It allows a viewer to develop relationships and patterns with geography to solve problems. A pattern is discovered much more quickly when looking a map rather than reading a lengthy text document. Hence, once GIS was brought into education, core concepts emerged through spatial relationships. GIS data can be viewed via a database, map, or model. Thus, mapping is only one aspect of the tool. In education, I teach Arcmap as well as other GIS tools. In Arcmap, users can analyze and edit attributes and physical land features on the surface of the earth. The ability to use Arcmap allows students to communicate about subject matter much more effectively. Bringing GIS into education not only allows students to communicate on subjects in a more effective manner, but it exposes them to an ever-growing job market that is requiring GIS skills. I have noticed that teachers who come from industry have a different perspective on managing the classroom. For myself, I often feel students are my employees, and I am training them to be industry professionals. My expectations are high.

Although I am considered a "rookie" educator who has to learn many educational practices by attending lectures and taking notes, my most meaningful experiences have come from hands-on experiences. Yes, teaching itself can be thought of as project-based or hands-on learning. Compared to industry, education provides the opportunity for providing leadership and positive societal impact to tomorrow's generation. Without a doubt, STEM PBL provide invaluable life lessons that can be used throughout their lives in solving problems, identifying variables, and formulating plans. STEM PBL provides for creativity, shared learning experiences, growth, and immediate reinforcement for both students and educators.

STEM PBL is a unique tool and teaching and learning opportunity. Experiences in STEM PBL bring to the surface the realization that without science, technology, engineering, and mathematics, the world, as we know it, would not function. If educators carefully examine what they do every day, they will find science,

R. M. Capraro, M. M. Capraro, J. Morgan and J. Scheurich, (eds.), A Companion to Interdisciplinary STEM Project-Based Learning: For Teachers by Teachers, 237–240.

technology, and mathematics. How can caring teachers express this in their lessons? What does this reliance look like in a classroom setting? Collaboration and exploring the fundamental concepts of what is really being taught is a start. STEM increases awareness of higher-order thinking through problem-solving applications that involve a collaboration of the STEM subject areas. Often in the hallway at school, other teachers and I think of lessons that incorporate both of our subject areas. For example, students must create a city. As they focus on composite shapes, surface area, parallel lines, vertical angles, height, and shadow, they use mathematics. Addressing surface water runoff, the main mechanism for erosion, we can produce runoff algorithms, curves, and predictive models dependant on rainfall. The project could satisfy the part of the social science curriculum that covers people traits and population concepts or pollution. The chemistry teacher could go into what comprises tap water and differentiate the harmful from the harmless. In history, students could create historical figures and conflicts in their city that match the traits of people and countries they are studying in history. Keeping it simple is key. Start small; start with one or two teachers and move to a larger collaborative group once you get the application of the idea firmed up. Creation and execution of the learning platform provide growth and opportunity in a STEM environment.

STEM FITS INTO A WELL-ROUNDED EDUCATION EXPERIENCE

Collaboration and communication should be the backbone of in-depth content study. Time needs to be dedicated to open dialogue and discussions of what is going to be taught in future weeks. Collaborative STEM PBL time should be separate from meetings or planning time, which are usually dedicated to housekeeping, grading, or parent conferences. Collaborative STEM PBL time should be a creative moment for brainstorming fun and interesting ideas for implementation. Having the ideas are great, but bringing them to life is what STEM PBL is all about. Some educators become overwhelmed by the engineering component of STEM and trying to integrate the design process. It is important to consider *ALL* components of STEM (science, technology, and math) and see that these components are the tools and gears for engineering. In addition, I find it helpful to think of engineering as a verb. Engineering is problem-solving, building, and creating. In a sense, most everything we interact with was engineered and many of us engineer every day. If students are constructing a city for a project, engineering is the physical creation of the city and its components as well as the creative thought process. Of course, professional engineers must have a stronger understanding of the fundamental concepts behind what they are creating. They need to have strong foundations in physics, technical writing, chemistry, and how these impact people. STEM education moves beyond rockets, cars, and bridges to create a philosophy and concept of problem-solving that leads to a product. If educators can understand this conceptual view of engineering and realize that mathematics, science, and technology are the tools for achieving it, then a whole new world of opportunities will open up for students. With STEM PBL, students are able to problem solve and physically see the product of their efforts. When this opportunity is created, students tend to latch on and create meaningful experiences.

The question is often asked, how to make STEM PBL successful? Again, support, communication, and collaboration are key. Without the administration's support, it might be a difficult task. Core (science, mathematics, language arts, and social studies) teachers are likely to have state-mandated curriculum and common assessments and assignments, which are typically comprised of worksheets and multiple-choice tests. If there is no way to get around that, use the elective teachers whenever possible. I am always excited and willing to change things up, visit classrooms, invite other classes to my classroom, and alter my own lesson plans. Accordingly, support and buy-in from all stakeholders is the foundation for "engineering" STEM PBL into a well-rounded education.

When we first adopted STEM PBL, the original STEM teachers at my school and I attended several workshops that expanded on theory and hands-on projects. These conferences excited me and provided direction. However, when new teachers entered the academy (the school where I work), they did not have the opportunity to experience the same inspiring conferences; consequently, they felt lost initially. Communication, as in any business, is vital to success in education. True transparency is helpful when building curriculum. As in any group situation, levels of communication will differ. Some concrete communication factors include: laying out each other's scope and sequence, looking at and dissecting the Texas Essential Knowledge and Skills (TEKS), open dialogue during collaboration time, and short hallway interactions.

Collaboration works side-by-side with communication. The more frequently teachers meet, the more their ideas will flow, i.e., brainstorming is an integral component for teachers as well as for students. Collaboration with students is equally important. I often rely on my students for feedback on how to improve projects, what topics to cover, and what concepts cause them to struggle. Feedback from the recipients creates a continuum for ongoing improvement of the "work in process." When there is an area in math a student has not mastered, most of the time, other students have not mastered it too. I can use that as an opportunity to focus my projects on some of the concepts with which they struggle. For example, one of those common concepts is measurement. I often try to include measurement elements in multiple projects. During teacher collaboration, different subject-matter experts should sit with one another. English teachers should not be paired with other English teachers, nor science teachers with science teachers. From my experience, I know that STEM education requires sound language arts skills and foundations in the social studies. However, English teachers often feel left out, or left to implement technical papers with little creativity. This can be solved through examining the science and mathematics in assigned English literature books, or working on fictional stories that express concrete mathematical and scientific concepts. Think about the basic ideas for English, analyze assigned stories, and some form of science and math can be found. STEM PBL is really a simple concept. It is human nature to resist change and overcomplicate new and unfamiliar territory. Collaboration is essential and over time will become easier; just remember to start small. Begin exploring different concepts to cover STEM PBL in your own classroom. Communicate what you have done to other teachers as the opportunity presents itself. Excitement is contagious and will grow. Eventually, lessons will fit entire themes and expand into several teachers' classrooms.

STEM PBL EDUCATION INTERESTS STUDENTS

Building and implementing (or should I say engineering?) STEM PBL is a challenge, and it takes a collaborative effort. Speaking from my own experiences with implementing STEM PBL, in order to build successful experiences within a school, a culture must be built. Creating that culture within a school can be very challenging especially for in urban, high poverty centers. Students will not engage in the learning process if they do not feel connected and if it differs from what they believe school to be like. Building a culture is affected by many factors including but not limited to administrative support, consistency, willing teachers, and the ability to modify state curriculum to facilitate designing projects to fit existing curriculum. I have had the fortunate opportunity to write my own curriculum; thus, STEM PBL was a natural process for me both as a teacher and as the person who was designing the custom learning experiences for my students. However, I understand that "Core" teachers may have difficulty implementing STEM PBL and may feel torn between what they want to do and what they think they "must" do. However, the goal is to improve student learning and meeting this goal will ensure that students make progress based on the identified benchmarks.

Furthermore, several students who have taken my classes reflect on the fact that the content presented can often be answered in more than one way. Thus, students have an opportunity to apply what they have previously learned in life to the real-world situations that are presented in class. My students have expressed appreciation about several aspects of the projects we have done in class. They enjoyed the group work, the hands-on components, and the learning about the various software programs, Word®, Excel®, PowerPoint®, Computer Aided Design, and various global information systems software packages. They were most appreciative that I did not explain every detail and let them draw upon their own skill set in order to solve the problems they faced in the STEM PBL lessons. Finally, they identified that the well-defined outcomes were instrumental in helping them refine their work, for example, "Do you think the library is conducive to learning?" When teachers pose problems in the form of questions students are overjoyed and motivated to investigate solutions.

I often present questions or ideas and let the students decide the process for solving them. During the first few projects, several students asked for assistance or to be hand-fed solutions. However, once they have completed a project, solved a problem, and viewed their own work/product, they thrive on being able to solve problems on their own. Students have said that PBL opens up the window of creativity and allows them to make an individual mark on their work/product. There is typically more than one way to answer a question or complete a project, just as in dealing with life's problems. Consequently, each student has something to bring to the table. Education becomes meaningful when students feel confident enough to mix their prior knowledge with their newly found knowledge to resolve issues and when students are passionate about what they are doing. When students and teachers find a passion for what they are doing, learning comes naturally. Likewise, when people truly enjoy what they are studying, learning, or teaching, engaging in tasks feels good.

A FINAL NOTE

STEM PBL should complement the existing curriculum. Do not reinvent the wheel; make the wheel better. Teachers should ask themselves, "Can the student apply what has been taught to solve real-world situations and do so utilizing meaningful knowledge?" If the answer is yes, then STEM PBL can most likely be a part of the teacher's routine.

The U.S. education system shows a significant amount of achievement gaps among gender and ethnicity. These gaps are exaggerated in urban schools with disproportionately lower achievement. STEM PBL aims to close these gaps by allowing students to test concepts with their hands and use their knowledge cumulatively. The hands-on approach introduces physical connections that helps students remember and apply content knowledge. If you ask students what they learned after a day of lecture some may reply with an in-depth response, most would say, "I don't know." However, with STEM PBL the responses will be varied but most will have a strong positive response. Any student can learn; it is the teacher's approach and style for developing the learning that opens up the door to student success. STEM content is creating a buzz all over the United States. Due to a lack of college-bound students majoring in engineering and mathematics, it is imperative that United States schools promote STEM education to prevent the potential weakening of our nation as a whole.

Who are the future American innovators and inventors? They are students; many of whom currently may not feel confident enough to apply what they have learned. Many are waiting for a connection to be made; STEM PBL can make that connection. Together, educators, administrators, legislators, students, and parents must seek higher-education innovation and solutions. The current education system needs improvement and innovative approaches to enhance the school day. I often laugh at the mouse-to-cheese behavior of some students. Some are so worried about the bell ringing that they pack up early and gather at the door. In a PBL environment, students work past the bell, and I am often writing them passes to their next class because they were so engrossed in their activities.

Models of innovation are among us and STEM PBL is one of them. The future is going to be different and ever-more changing and so should the education we provide today's child. The education arena today needs to move forward and the perpetual motion of a modernizing education should be taken seriously. Educators need to take charge of education and have a stake in a model of progress for improvement of the American education system.

LaVergne, TN USA
28 April 2010
180948LV00001B/19/P